04-CTS-1

MINIMIZATION OF ODORS AND CORROSION IN COLLECTION SYSTEMS PHASE 1

Dirk Apgar
King County, Washington State

Jay Witherspoon
CH2M HILL

Chris Easter
CH2M HILL

Shabir Bassrai, Carla Dillon, Ed Torres
Orange County Sanitation District

Richard Corsi, Ph.D.
University of Texas

Bob Forbes, Chris Quigley, Matthew Ward
CH2M HILL

Robert Morton, Jeff Weiss
Los Angeles County Sanitation Districts

Robert P. G. Bowker
Bowker and Associates

Steve Davidson, Philip Wolstenholme
Brown and Caldwell

Jim Joyce
Odor and Corrosion Technology Consultants

Richard Stuetz, Ph.D.
University of New South Wales

2007

WERF

The Water Environment Research Foundation, a not-for-profit organization, funds and manages water quality research for its subscribers through a diverse public-private partnership between municipal utilities, corporations, academia, industry, and the federal government. WERF subscribers include municipal and regional water and wastewater utilities, industrial corporations, environmental engineering firms, and others that share a commitment to cost-effective water quality solutions. WERF is dedicated to advancing science and technology addressing water quality issues as they impact water resources, the atmosphere, the lands, and quality of life.

For more information, contact:
Water Environment Research Foundation
635 Slaters Lane, Suite 300
Alexandria, VA 22314-1177
Tel: (703) 684-2470
Fax: (703) 299-0742
www.werf.org
werf@werf.org

This report was co-published by the following organizations. For non-subscriber sales information, contact:

IWA Publishing
Alliance House, 12 Caxton Street
London SW1H 0QS, United Kingdom
Tel: +44 (0) 20 7654 5500
Fax: +44 (0) 20 7654 5555
www.iwapublishing.com
publications@iwap.co.uk

Library of Congress Catalog Card Number: 2007929714
IWAP ISBN: 978-1-84339-791-5

King County Wastewater Treatment Division (Washington), CH2M HILL

The research on which this report is based was developed, in part, by the United States Environmental Protection Agency (EPA) through Cooperative Agreement No. CR-831559-01 with the Water Environment Research Foundation (WERF). However, the views expressed in this document are solely those of King County Wastewater Treatment Division (Washington) and CH2M HILL and neither EPA nor WERF endorses any products or commercial services mentioned in this publication. This report is a publication of WERF, not EPA. Funds awarded under the Cooperative Agreement cited above were not used for editorial services, reproduction, printing, or distribution.

ACKNOWLEDGMENTS

The research team very much appreciates the efforts of the various utility partners and support of both WERF member and nonmember utilities that contributed to this research effort. The team also expresses its thanks for those agencies and vendors who responded to the researcher's request for gray (unpublished) literature related to Odor and Corrosion in Collection Systems. They appreciate the assistance of the 66 agencies, firms, and vendors who contributed their work and efforts by furnishing this information for their research efforts.

Report Preparation

Principal Investigators:
Dirk Apgar, M.S., P.E.
King County Wastewater Treatment Division, Washington

Jay Witherspoon, M.S.
CH2M HILL, Bellevue, Washington

Project Team:
Shabir Bassrai
Orange County Sanitation District

Bob Bowker, M.S.
Bowker and Associates, Inc., Maine

Richard Corsi, Ph.D.
University of Texas

Steve Davidson
Brown and Caldwell

Carla Dillon
Orange County Sanitation District

Chris Easter, M.S.
CH2M HILL, Richmond, Virginia

Bob Forbes
CH2M HILL, Raleigh, North Carolina

Jim Joyce
Odor and Corrosion Technology Consultants

Robert Morton, M.S.
Los Angeles County Sanitation Districts

Chris Quigley
CH2M HILL, Austin, Texas

Richard Stuetz, Ph.D.
University of South Wales

Ed Torres, M.S.
Orange County Sanitation District

Matthew Ward, M.S.
CH2M HILL, Austin, Texas

Jeff Weiss
Los Angeles County Sanitation Districts

Philip Wolstenholme
Brown and Caldwell

Project Subcommittee

Michael W. Sweeney, Ph.D., RC Liaison
Principal Consultant, EMA, Inc.

Gail Chesler, Ph.D., P.E.
Central Contra Costa Sanitary District, Martinez, California

Stephen R. Harper, Ph.D., P.E.
Senior Technical Associate, O'Brien and Gere Engineers, Alpharetta, Georgia

Richard J. Pope, P.E.
Malcolm Pirnie, White Plains, New York

Bradley Striebig, Ph.D.
Assistant Professor of Civil Engineering, Gonzaga University, Spokane, Washington

James D. Prah, Ph.D.
U.S. EPA Human Studies Facility, Chapel Hill, North Carolina

Water Environment Research Foundation Staff

Director of Research: Daniel M. Woltering, Ph.D.
Senior Program Director: Amit Pramanik, Ph.D.

ABSTRACT AND BENEFITS

Abstract:

In this Phase 1 report, the researchers, from academic, utility management, and consulting communities, summarize the current knowledge of odor and corrosion in collection systems culled from a literature survey of over 4,000 published abstracts and gray literature sources supplied by utilities, manufacturers, and vendors in an effort to better explain the science underlying odor generation, sewer ventilation, odor characterization and monitoring, and corrosion mechanisms. An easily understandable, "plain-English" overview of odor and corrosion in collection systems, including how odor- and corrosion-causing compounds are formed and what to do to control them, is presented. The overview of odor- and corrosion-causing compounds includes formation mechanisms and odor-sampling methods and identifies knowledge gaps in the literature, such as the need to link sensory perceptions of odorants to their chemical composition. Discussions of gas phase and liquid phase odor and corrosion is followed by an identification of the top research gaps in the literature: collection system air movement and ventilation; the relationship of gas phase H_2S to concrete corrosion rates; and non-H_2S-related odor compounds and the effectiveness of various gas and liquid phase treatment technologies for them.

Benefits:

- Provides easily understandable overview of odor- and corrosion-causing compounds.
- Summarizes the latest knowledge reported in the literature.
- Provides unified approach to liquid and gas phase preventions.
- Combines expertise regarding utility and asset management, affordability, regulatory and permitting, public outreach, characterization, assessment and measurement, and system planning, design, odor and corrosion abatement, and operation and maintenance.

Keywords: Collection system, odors, corrosion, hydrogen sulfide.

TABLE OF CONTENTS

LIST OF TABLES

LIST OF FIGURES

LIST OF ACRONYMS

BOD	biochemical oxygen demand
CFD	computational fluid dynamics
cfm	cubic feet per minute
CIPP	cured-in-place pipe
COD	chemical oxygen demand
D/T	dilutions-to-threshold
DO	dissolved oxygen
EBRT	empty-bed residence time
ECIS	east central interceptor sewer
FOG	fats, oils, and grease
FSC	flameless sulfur chemiluminescence
GC/FID	gas chromatography with flame ionization detection
GC/FPD	gas chromatography with flame photometric detection
GC/MS	gas chromatography with mass spectrometry
GC/MS/O	gas chromatograph mass spectrometer olfactometer
GC/NPD	gas chromatography with nitrogen-phosphorus detection
HS-SPME	headspace solid phase microextraction
LACSD	Los Angeles County (Calif.) Sanitation District
LCA	life-cycle analysis
LEL	low explosive limit
MBR	membrane bioreactor
MIC	microbially induced corrosion
OCSD	Orange County (Calif.) Sanitation District
OEC	odor emission capacity
OP	odor potential
ORP	oxidation-reduction potential
OU	odor unit
PDMS	polydimethylsiloxane
POTWs	publicly owned treatment works
ppb	parts per billion
ppm	parts per million
PSA	pressure swing adsorption
QA/QC	quality assurance/quality control
SPACA	septicity prediction and control algorithms
SPME	solid phase microextraction
SRB	sulfate-reducing bacteria
TON	threshold odor number
U.S. EPA	U.S. Environmental Protection Agency
VFA	volatile fatty acid
VOC	volatile organic compound
WATS	wastewater aerobic–anaerobic transformations in sewers
WERF	Water Environment Research Foundation

EXECUTIVE SUMMARY

Odor complaints, corrosion, and costly sewer rehabilitation projects are damaging public relations, degrading wastewater facilities, and stressing already-limited collection system budgets. Research has revealed several seemingly unrelated factors that in combination cause the increased collection system odor and corrosion problems that municipalities face today. Significant gaps exist in our understanding of odor production and corrosion in collection systems. As a result, odor and corrosion prevention in collection systems is as much art as science.

In response to these issues, the Water Environment Research Foundation (WERF) selected King County (Wash.) Department of Natural Resources and Parks—Wastewater Treatment Division and CH2M HILL to conduct a two-year project focusing on minimizing and preventing odors from, and corrosion of, wastewater collection systems. The research team includes practicing wastewater collections systems experts from major university research institutions, highly specialized and qualified consulting firms, and numerous collection system owners and operators. The two-year project is divided into three phases: 1) literature assessment; 2) research agenda implementation; and 3) practitioner's guide.

This Phase 1 report presents the results of a comprehensive literature review and assessment. It focuses on the odor and corrosion approaches, strategies, challenges, and successes and failures as described in both the open peer-reviewed and gray literature. The team's efforts to compile the database included information-sharing partnerships with municipal utilities, the academic community, and the profession, all on a global basis. Current literature on wastewater collection system odor and corrosion was identified and reviewed systematically, and each paper's essential information was documented electronically.

The published literature search generated 2,945 references, of which 263 were deemed worthy of inclusion in the database. These papers were mostly conference papers and journal articles published after 2000. The gray literature search generated 52 papers from nine organizations. Of these, 33 contained information relevant to the project and were included in the database. From a previous WERF study 317 papers were screened (published and gray) and 85 were included in the review. The quality control process generated an additional 44 papers identified by the project team. A Googlescholar search produced 485 references, 14 of which were worthy of inclusion in the database, but had not yet been identified. The total number of references screened was 3,842, with a total of 435 papers included in the database. Nineteen of these papers were printed in a foreign language and were not translated, and four could not be obtained. The total number of papers included in the database was 412.

A "plain-English" guide gives a useful and easily understandable overview of odor- and corrosion-causing compounds in collection systems, including how these compounds form, how they are measured, and how they can be controlled. As well as being a comprehensive overview for the layperson, the plain English guide introduces the pertinent topics for the wastewater professional and the literature search findings discussed in more detail in the chapters that follow.

First, in Chapter 4.0 the team describes advances in identifying malodorous compounds emitted from sewers, biological mechanisms through which odorous and corrosive chemicals form in sewers, and emerging measurement methods. Insights into state-of-the-art methods for measuring and preventing the formation of odor- and corrosion-causing compounds are presented. For example, the researchers observed that past studies related to corrosion in sewers have focused almost entirely on the production and release of hydrogen sulfide and formation of sulfuric acid that leads to corrosion of sewer pipes. Additional research related to whether other biochemical reaction products in sewers lead to corrosion, or enhance the corrosion related to hydrogen sulfide releases, is warranted. It was concluded that fermentation processes and volatile fatty acid formation remain areas of uncertainty and are worthy of additional basic (laboratory) research. Furthermore, the concept of odor potential should be expanded to include not only the current state of wastewater but also the potential to form odorants following chemical and biological transformations. This necessarily requires greater knowledge of parameters that lead to transformations and the transformation pathways themselves, and thus a need for more laboratory research. Additional field studies are needed to link sensory perceptions of odor complaints, that is, annoyance, with the chemical composition of odorants. Olfactory gas chromatography with mass spectrometry enables odorants in gas samples to be separated and identified individually and allows for the odor contribution of each compound to be characterized and quantified using an odor port. Finally, two promising technologies that might facilitate the identification of odorous gases and/or nature of odors include solid phase micro-extraction for the concentration of trace concentrations of odorous compounds and electronic nose systems. However, each of these technologies requires further laboratory research, and ultimately field evaluations, to optimize

Next, in Chapter 5.0, the team explains recent advances in modeling wastewater collection system odors and corrosion. Six models are reviewed and evaluated, and their strengths and weaknesses noted. Each is examined for how well it addresses the multiple biological, chemical, and physical mechanisms and processes that cause odors or corrosion in collection systems. These mechanisms/processes include:

- Liquid phase sulfide generation
- Liquid-to-vapor H_2S mass transfer
- Natural liquid phase chemical and biological oxidation of sulfides
- Dissolved oxygen impacts on sulfide generation
- Liquid and vapor phase bulk transport of sulfides and H_2S, respectively
- Vapor phase H_2S-induced concrete corrosion

Several wastewater collection system models have been developed to facilitate the assessment of sulfide generation and potential release into the sewer headspace atmosphere. While several of the models rely on similar methods for approximating sulfide generation and liquid-to-vapor mass transfer, the methods for predicting ventilation rates and inclusion of concrete corrosion as an H_2S loss term are not consistent. An improved understanding of the fundamental principles associated with these two areas—natural ventilation of sewers and quantification of vapor phase H_2S loss associated with concrete corrosion—are therefore considered key areas for future research.

The existing knowledge base related to sewer ventilation is described in Chapter 6.0. The nature and extent of air exchange (ventilation) between the ambient atmosphere and headspace of sewers may be the most important variable affecting the generation and release of odorous and corrosive gases in gravity-flow sewers. Sewer ventilation is required to replenish the oxygen content in the headspace of sewers, and thus affects the dissolved oxygen content of wastewater. It can also affect the degree of condensation on sewer crowns (and hence sewer corrosion), and it defines the locations where odorous gases are released from the sewer system.

Despite the obvious importance of sewer ventilation to collection system odor and corrosion problems, there has been little progress related to a fundamental understanding of the competing factors that affect ventilation. Factors required to relate changes in wastewater flow, headspace above the liquid, or physical structure of sewer systems to sewer ventilation are needed. Improved knowledge in this area is critical to advancing integrated solutions to sewer odor and corrosion problems. Research has tended to evaluate specific collection systems and focus on solving site-specific odor problems. Full-scale field data often vary considerably from predicted values developed from laboratory pilot tests. None of the previous work attempts to define sewer headspace ventilation in terms similar to airflow in a duct and then relate those parameters to wastewater flow and interceptor characteristics. Furthermore, collection system leak-tightness, while being a critical ventilation parameter, has not been assessed in any of the work thus far. A specific research agenda related to sewer ventilation is described at the end of Chapter 6.0 and in the research gap summary Chapter 10.0.

Chapter 7.0 includes present current, successful, state-of-the-art techniques regarding liquid phase odor and corrosion prevention and provides information on how they work, mechanisms and key equations, and how they are being used globally. A proven approach to reducing both odor and corrosion inside the collection system requires a cost-effectiveness calculation and present worth value investigation before settling on an approach or implementing one in the collection system. Research gaps are provided for future research consideration. These gaps include: Life cycle analysis of collection system chemical addition impacts on downstream processes, pure oxygen treatment of gravity sewers, magnesium hydroxide odor control mechanisms other than pH adjustment, and liquid phase biological treatment. Details related to each knowledge gap are provided at the end of Chapter 7.0.

Traditional and innovative gas phase odor treatment approaches and equipment are presented in Chapter 8.0, with a focus on pump stations and other wastewater collection system locations or structures that could potentially emit odors into the surrounding atmosphere. Traditional and innovative approaches are explored for use at the various odor-emitting processes or structures, including chemical and carbon scrubbers, membrane reactors, biotowers, and biofilters. Innovative approaches include membrane reactors and biotechnology as well as combining traditional approaches with new ones to meet very high odor removal needs. Much information on these successful odor control approaches and applications is available. Insights on how and where to use these gas phase odor treatment systems throughout the collection system are provided. Specific research needed to fill current knowledge gaps include: identification of compound in sewer gas that lead to odor complaints, effectiveness of treatment systems for odorous gases other than H_2S, and effectiveness of biotechnology in cold climates.

Also presented, in Chapter 9.0, are corrosion mechanisms that have been described in the literature only recently; measures to control such mechanisms have been developed in parallel. Three such measures include adding coatings, most frequently on the crown of the pipe; lining the pipe wall; and installing corrosion-resistant pipe material. The many methods currently available to repair, rehabilitate, and replace wastewater collection systems are examined. Each method's appropriate uses and limitations are discussed, and the suitability of a given technology for a specific project is provided. This chapter also includes a discussion of chemical addition and ventilation approaches as means to protect and prevent corrosion forming compounds in the gas phase of the collection system. It was concluded that guidance is needed for agencies needing to provide pipeline corrosion protection for both new and existing installations. While many vendors may claim that their corrosion protection techniques are the most effective, objective data are needed. Pipe material corrosion resistance techniques should be compared to other techniques such as chemical addition.

The literature search and evaluation initially identified 13 significant gaps in research and knowledge and suggests potential research to be undertaken to fill each knowledge gap. Table ES-1 summarizes those gaps and the projected costs of research for most of these gaps:

Table ES-1. Research Gaps and Their Projected Costs.

Research Gap	Projected Cost ($)
1. Air movement and displacement in collection systems	212,731
2. Odorous and corrosive compounds found in collection systems	163,000
3. Effective treatment of non-H2S odorous compounds	151,900
4. Relationship between H_2S gas concentrations and corrosion rates	196,985
5. Impacts of collection systems' chemical dosing on the downstream treatment plant	100,000
6. Sulfide generation rates and the importance of substrates	117,000
7. The practicality and related costs of using oxygen treatment in gravity sewers	118,000
8. Biological treatment using enzymes and bacterial cultures and the effectiveness of these approaches	52,000
9. The mechanisms of magnesium hydroxide pH control	88,000
10. Design guidance for sizing air jumpers and related ventilation systems (may become part of the effort associated with Gap 1)	36,000
11. The effectiveness of biotechnology-based odor control systems in cold climates	143,000
12. Odor and corrosion impacts from fats, oils, and grease	75,000
13. The odor and corrosion effect of using sewers as reactor vessels to achieve partial wastewater treatment	192,000

Note: Cost is initial estimated cost including in-kind contributions. Final cost would be developed for higher priority research gaps that move forward.

A briefing held with the PSC allowed the initial list of 13 research gaps to be focused down to the six highest-priority gaps. A WERF subscriber survey was then completed in order to further refine this list and select the highest priority research gaps. Four top-priority research gaps were identified on the basis of this survey and review with WERF:

- Improving understanding of the nature of sewer system ventilation
- Improving understanding of the relationship between H_2S gas phase concentrations and corrosion
- Better defining the range of odor- and corrosion-causing compounds that come from collection systems (a subset of this topic being the desirability to better understand the impacts of fats, oils, and grease on odor- and corrosion-causing compounds in collection systems receiving high doses of fats, oils, and grease)
- Determining the effectiveness of liquid and gas phase treatment approaches for non-H_2S Odor compounds

On the basis of input and guidance from WERF, WERF subscribers, and utility owners the highest priority research gap areas will be pursued in following phase of this study within the limits of available scope and funding or will be pursued through collaborative research programs.

CHAPTER 1.0

INTRODUCTION

1.1 Overview

In May 2005, the Water Environment Research Foundation (WERF) selected a research team led by the King County (Washington) Department of Natural Resources and Parks—Wastewater Treatment Division and CH2M HILL to perform WERF Project 04-CTS-1, "Minimization of Odors and Related Corrosion in Collection Systems." This team has been assisted by several consulting firms, major research universities, and publicly owned treatment works (POTWs). In addition, this first phase report has been reviewed by a technical advisory committee of experts and regulatory specialists in the collection system odor-related field and by several POTWs.

The multi-year project is divided into three phases: Phase 1, Literature Assessment; Phase 2, Research Agenda Implementation; and Phase 3, Practitioners Guide. This Phase 1 report presents the results of comprehensive literature assessment conducted by the team.

The authors of this report have been involved in a comprehensive odor and corrosion control and prevention research project that seeks to identify and explore gaps in knowledge concerning odor and corrosion issues that affect the wastewater collection system. Participating consultants, universities, and POTWs have provided more than $1 million of in-kind contributions of time and services for this project.

This report focuses on the odor and corrosion approaches, strategies, success and failure stories, and challenges discovered through a comprehensive literature search. The researchers' efforts to compile the database included information-sharing partnerships with municipal utilities, the academic community, and the profession, all on a global basis. This is the start of a central clearing house related to wastewater collection system odors and corrosion.

1.2 Phase 1: Literature Assessment and Review on Collection System Odors and Corrosion

The team has completed Phase 1. The project team completed a comprehensive review of published literature and unpublished documents within the wastewater collection system industry. The project team examined papers and reports associated with nearly all wastewater collection system processes and infrastructure, including pipes, trunks, force and rising mains, pump stations, junction boxes and other related processes or infrastructure. In addition, the project team reviewed "gray literature" describing the experiences of utilities in the wastewater collection industry with respect to odor and corrosion assessment, measurement, modeling, controls and prevention.

The research team pulled together existing collection system odor and corrosion knowledge seen in 17 research databases, including information from 33 gray literature sources supplied by utilities, manufacturers, and vendors, and a total of 435 research papers screened from nearly 4,000 abstracts.

1.3 Objective

The objective of this report's authors and contributors is to transfer state-of-the-art technology and information gained from the literature survey to the collection system owner and designer on odor and corrosion assessment, measurement, characterization, monitoring, and prevention. The general tasks of this assessment were to summarize the contemporary literature in wastewater collection system disciplines to:

- Improve odor and corrosion characterization for wastewater collection systems
- Determine applicable odor and corrosion assessment models, sampling, and measurement approaches for potential sources for characterization purposes
- Define state-of-the-art odor and corrosion control and prevention technologies, additions, and strategies
- Compile lessons learned for viable and innovative odor and corrosion control and prevention strategies and technologies into a "plain English" guide for owners, designers, and operators
- Identify key gaps in knowledge of technology or infrastructure in addressing odor and corrosion issues form the entire collection system.

1.4 Background

Wastewater collection system odors and corrosion issues continue to grow in importance to the community and to system owners and operators. Owners and operators daily address a number of planning, operations, maintenance, odor, corrosion, and regulatory compliance issues to provide cost-effective, environmentally sound, and reliable wastewater collection services to their customers. Collection systems are defined as the conveyance of wastewater through a series of pumps, pipes, trenches, tunnels, and by other means to a wastewater treatment system.

Many collection systems are aging faster than planned due to "odor compound"-induced corrosion, which is typically caused by odorous gases that are generated or released inside the collection system. Collection systems also have a high potential for odors that impact surrounding residents, businesses, and communities. Plus, the collection system is located throughout most communities, cities, and towns, as well as directly connected to most houses, businesses, parks, and other sources of wastewater discharges.

The field of odor and corrosion prevention and control in collection systems is not new, but it is ever-changing. Odor and corrosion prevention in collection systems is as much art as science. Common control methods are selected based on practical experience as opposed to a fundamental understanding of why and when methods will be successful. Although much is known regarding the cause of odorous gases in the collection system, the underlying science and mechanisms of odor generation, sewer ventilation, odor characterization and monitoring, and corrosion mechanisms need further research. The cost of preventing and controlling odors and corrosion is continuously increasing as collection systems age and outside sources of funding become scarcer. Numerous approaches to odor and corrosion prevention in control in collection systems have been tried over time with varying degrees of success. There is no central clearing house for ideas or successes and failures.

1.5 Participants

The project team comprised administrative staff and technical specialists from King County, Wash.; the Los Angeles County (Calif.) Sanitation District (LACSD); Orange County (Calif.) Sanitation District (OCSD); CH2M HILL; the University of Texas at Austin; University of New South Wales, Australia; Odor and Corrosion Technology Consultants; Brown and Caldwell; Environmental Management Consulting; and many others. Oversight was provided by the technical advisory committee, a WERF project subcommittee (see Acknowledgments), and many POTWs, listed below:

- Butler County Department of Environmental Services, Oh.
- Clayton County Water Authority, Ga.
- Clayton County, Ga.
- Dublin San Ramon Service District, Ca.
- East Bay Municipal Utility District, Oakland, Ca.
- Gwinnett County, Ga.
- Hampton Roads Sanitation District, Va.
- Little Blue Valley Sewer District, Mo.
- Los Angeles Sanitation District, Ca.
- Melbourne Water, Melbourne, Australia
- Metro Wastewater Reclamation District, Denver, Co.
- Metropolitan Water Reclamation District of Greater Chicago, Il.
- Northumbrian Water, Durham, U.K.
- Orange County Sanitation District, Ca.
- Philadelphia Water Department, Pa.
- San Antonio Water System, Tx.
- Stuart, Fl.
- United Water Services, Tx.
- WaterCare Services, Ltd., Auckland, New Zealand
- West Palm Beach, Fl.

1.6 Organization

Information for this report was compiled by a team of experts who have been working and interacting in the field of wastewater collection system odor and corrosion prevention for many years. Subject matter by chapter is as follows:

Chapter 2.0 provides an easy-to-read overview of odor and corrosion issues in collection systems.

Chapter 3.0, Literature Search and Review, summarizes the comprehensive approach taken by the team to gather and review collection system odor and corrosion assessment, measurement, and prevention documents seen since 2000 in addition to information from a previous WERF literature search spanning 1990-2000 and grey literature since 1990.

Chapter 4.0, Odor- and Corrosion-Causing Compounds, reviews the literature findings related to what odor- and corrosion-causing compounds have been identified as well as discussing their formation mechanisms and methods for analysis of these compounds.

Chapter 5.0, Wastewater Collection System Modeling, reviews the literature on how odorous and corrosive gases are emitted and highlights key points about the various emissions models currently used in the industry.

Chapter 6.0, Sewer Ventilation, reviews the literature findings on air movement in collection systems and discusses key parameters that effect it.

Chapter 7.0, Liquid Phase Odor and Corrosion Control, reviews findings on a variety of liquid phase treatment options to reduce odor- and corrosion-causing compounds, such as including oxidation chemicals and iron salts, pH shifting, and nitrate addition, and using bacterial cultures and enzyme supplement products.

Chapter 8.0, Gas Phase Oder Treatment, reviews literature observations and findings on a variety of gas phase treatment options for collecting and treating odorous and corrosive headspace air in collection systems.

Chapter 9.0, Corrosion Mechanisms and Protection, reviews literature observations about how corrosion occurs and what measures can be take to mitigate corrosion, such as coatings and linings.

Chapter 10.0, Research Gaps Evaluation and Summary, summarizes all of the research knowledge gaps identified during the literature search and includes both a problem statement as well as a general plan for how the knowledge gaps might be investigated and filled.

CHAPTER 2.0

Plain English Guidance

2.1 Background

Problems associated with odors and corrosion in wastewater collection systems have become much more frequent in recent years from both public perception and infrastructure protection standpoints. Odor complaints, corrosion, and costly sewer rehabilitation projects are damaging public relations, degrading our wastewater facilities, and stressing already-limited collection system budgets. New research has revealed several seemingly unrelated factors that have combined to cause the increased collection system odor and corrosion problems we face today. This chapter reveals these factors and discusses their impact on collection system odor and corrosion. This chapter also unveils a new strategy for evaluating odor and corrosion control alternatives by interrupting the various physical, chemical and biological processes which produce hydrogen sulfide in sewers.

As odors have become more prevalent and corrosion more damaging, we find that many of the control strategies and technologies that worked well in the past cannot handle the increase, and many are no longer cost-effective. Recent advances in the understanding of odor and corrosion processes and the introduction of many new and innovative control methods and technologies necessitate this significant technical update. The information in this manual will be valuable for collection system managers, design engineers, operators, and others responsible for today's modern collection systems.

2.2 Introduction

The information presented in this manual has been authored by recognized experts in the field of wastewater odor and corrosion control and represents the most current information available. Technical details and case histories provide an in-depth analysis of real-world control alternatives applicable to modern collection systems. This manual's authors believe, however, that the reader should first be familiar with some of the basic scientific concepts related to collection system odor and corrosion in order to fully utilize this material. The authors have therefore prepared this chapter to present some of the basic science that later chapters build upon. This chapter also discusses the common biological and chemical principles that users of this manual will need to understand and apply the control measures discussed in later chapters.

2.3 Collection Systems Overview

As long as there have been sewers there have been sewer odors. The emission of "foul vapors" from sewer catchments and openings has been a common problem for developing civilizations throughout history. To compound this issue, sewers have demonstrated the unique ability to corrode, deteriorate, and sometimes collapse for seemingly unknown reasons. Over thc past 80 years researchers have identified the primary cause of sewer odor and corrosion as hydrogen sulfide gas, although many other odorous compounds are also found in sewers.

Hydrogen sulfide gas is the only substance produced in sewers that can cause damaging corrosion, and it is therefore provided some special attention in this chapter.

Recent studies have shown that hydrogen sulfide gas concentrations in sewers have been increasing since about 1985. While some municipalities have been affected more than others, owners and operators of all wastewater systems are now experiencing increasing odor and corrosion problems but are armed with outdated knowledge and technologies. This chapter presents the most recent developments concerning this increase in hydrogen sulfide, its role in odor and corrosion, and best methods for control.

Before we can control something we need to understand it. Something as simple and common as odor is no different. Collection system owners and operators must first know what "odors" are, some of their more important properties, and how they impact humans. We can then build a solid foundation for a sound, scientific approach to controls.

2.3.1 Odors in Sewers

Wherever you find them, odors are simply airborne (volatile) chemical compounds. When those volatile chemical compounds interact with certain protein receptors on cells in our nose, our brain registers an "odor." A well-known example of an odorous volatile chemical compound would be perfume, a volatile liquid that people apply to their bodies to make them smell more pleasant. Odors can play an important role in human psychology by directly affecting one's perception of their personal well being. When perfume molecules continuously release to the air, humans detect these molecules with their sense of smell, recognize them as pleasant and feel good about their environment. Volatile odor compounds found in collection systems are released and detected the same way; however, not many of them are likely to be recognized as pleasant.

The volatile odor compounds found in sewers generally have three basic origins: feces, chemical discharges into the system, and the products of biological and chemical reactions in sewers. Natural fecal odor compounds are produced in the human digestive system, dissolved into the wastewater, and released directly into the sewer airspace. Many other odorous compounds are discharged into sewers from factories, restaurants, industries, and rendering plants, for example.

The most important compounds related to odor and corrosion in sewers are generated in the sewer through biological and chemical interaction of the mixed sewage components. Odor compounds produced this way include hydrogen sulfide, which is a primary focus of this chapter.

2.3.2 Mechanisms of Sulfide Generation

The role of hydrogen sulfide in the sewer odor and corrosion process is so important that some time must be spent discussing how it is produced and how it reacts with other sewage components. As mentioned above, sulfide is generated in the wastewater collection system from the interaction of received wastes. Once formed, it can exist in one or more of several different chemical forms, each is important to the understanding of control strategies. The following is a brief step-by-step discussion of exactly how sulfide and hydrogen sulfide are formed in sewers.

Sulfide generation is a biological process that occurs only in the submerged portion of sanitary sewers. Fresh domestic sewage entering a wastewater collection system is usually free of sulfide. However, biological interactions in the wastewater quickly start producing sulfide. The primary sewer conditions which cause and accelerate sulfide generation are listed below.

1. High BOD_5 concentration (a measure of the strength of the wastewater)
2. High wastewater sulfate concentration
3. Elevated wastewater temperature
4. Long hydraulic detention time in the collection system
5. Many pumping stations and long force mains
6. Low dissolved oxygen concentration in the wastewater
7. Sewer debris

2.3.2.1 Slime Layer

The first step in the sulfide production process is the growth of a thin layer of biological slime below the water level in a gravity sewer or force main. This layer is composed of bacteria and inert solids held together by a gelatinous slime and covers virtually all underwater surfaces. When this biological slime becomes thick enough, the side next to the pipe wall becomes anaerobic. Approximately two weeks is required to establish a fully productive slime layer in new sewers. Once formed, slime layers become permanent fixtures in sewers; virtually all sewers have a slime layer. Although formed from many different bacterial species, many of the slime layer organisms are in a symbiotic relationship, providing shelter, food and energy for each other. By living in close proximity and sharing resources, slime layer bacteria thrive in this artificial ecosystem. Figure 2-1 shows a typical sewer slime layer, the generation of sulfide and the release of hydrogen sulfide gas.

As mentioned above, many different species of bacteria reside in the slime layer. Some species can only be found in one of the slime layer “zones” whereas other species can live in more than one. Their survival depends upon their ability to adapt and carry out life processes under the conditions in each zone. One of the most important conditions controlling where organisms live is the presence of oxygen or oxygen-containing compounds. The oxidation-reduction potential (ORP) of a particular zone can be measured in millivolts (mV) and is a measure of the amount and type of oxygen available. Anaerobically produced odor compounds are generally more offensive to humans than aerobically produced odors, a concept that is explained further in the next section. Therefore, odors (and hence corrosion) generally get worse as the ORP of a biological system drops. Figure 2-2 presents a typical biological ORP scale.

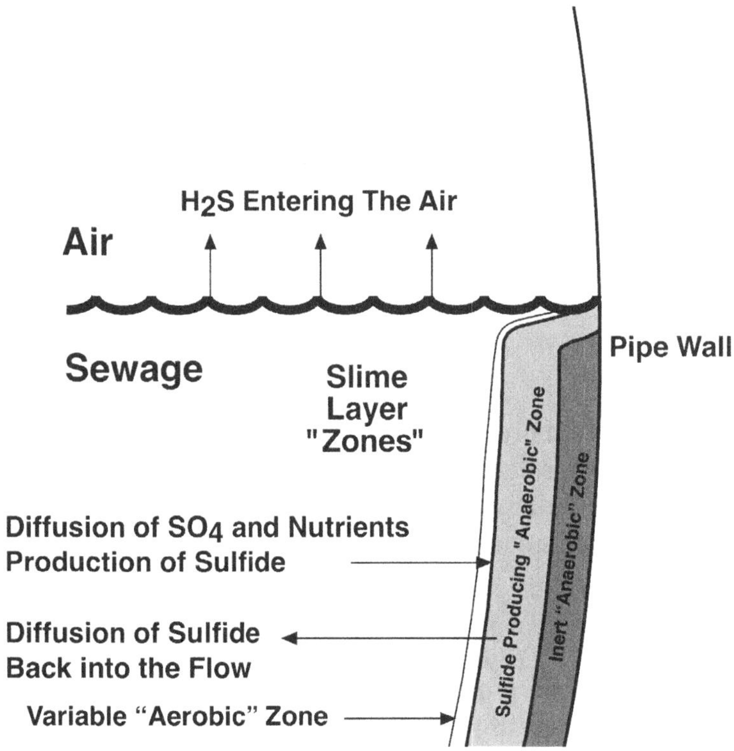

Figure 2-1. Slime Layer Chemistry and Biology Illustration.

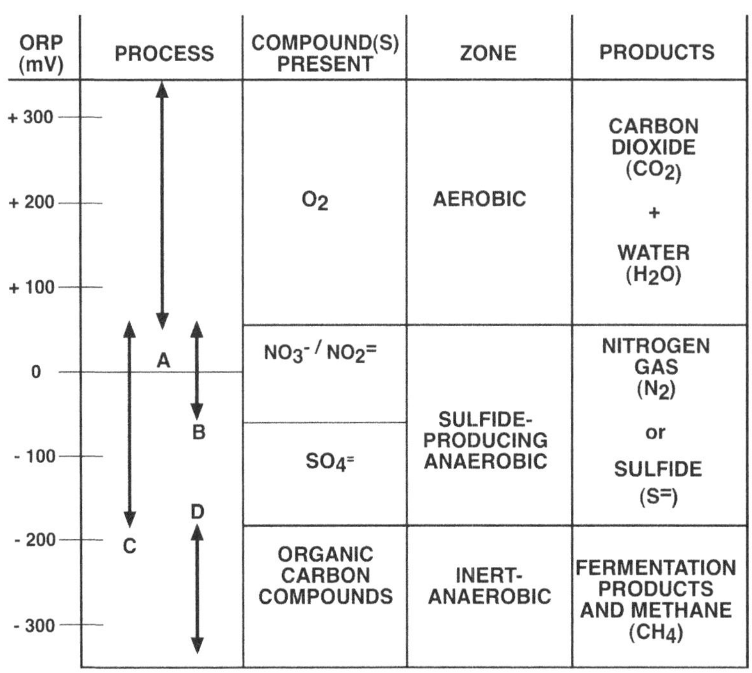

A = Organic Carbon Oxidation
B = Denitrification if Nitrate/Nitrite Present
C = Sulfate Reduction w/o Nitrate/Nitrite Present
D = Fermentation and Methane Generation

Figure 2-2. Biological ORP Scale.

The ORP values in Figure 2-2 are compiled from several references and represent an accurate averaging of variations which exist in the literature. The slime layer could be represented on this figure as having the highest (aerobic) ORP values at the outer surface nearest the wastewater and the lowest (anaerobic) values near the pipe wall.

As indicated in Figure 2-2, aerobic biological processes persist down to an ORP of around +50 millivolts (mV), at which point dissolved oxygen generally disappears. Between +50 and −200 is the range where only combined sources of oxygen are found. Bacteria under these conditions can use both nitrate and sulfate but prefer nitrate when present. It is this range of ORP values that dominates the slime layer biochemistry and is the primary cause of all sulfide-related odor and corrosion. Below −200 mV is the domain of the stick anaerobic and fermentative bacteria.

Many different species of bacteria reside in the slime layer. Some species can be found only in one zone, whereas other species can function in more than one. Their survival depends upon their ability to adapt and carry out life processes under the conditions in each zone.

2.3.2.2 Aerobic Zone

If the wastewater contains sufficient dissolved oxygen, the slime layer will be covered with an aerobic zone. The aerobic zone is characterized by the presence of dissolved oxygen (O_2), which penetrates a short distance into the biological slime. The bacteria that live in this zone use oxygen to break down organic compounds and carry out their life processes. Bacteria are able to get more energy from food sources when they use dissolved oxygen as opposed to combined sources of oxygen such as nitrate or sulfate. Some bacteria, called "strict aerobes," can use only oxygen; other bacteria, called "facultative bacteria," can use dissolved oxygen when it is present but can switch to combined sources of oxygen when O_2 is depleted. The primary byproducts of aerobic decomposition are carbon dioxide and water. The aerobic zone is usually very thin, if present at all.

The aerobic zone does not produce sulfide and is actually an active consumer of sulfide generated by the zone beneath it. When aerobic wastewater conditions permit the formation of an aerobic zone, the bacteria living there consume the sulfide and other odor compounds coming from below. Aerobic bacteria consider sulfide as a readily consumable food source and it is rapidly oxidized to sulfate and other harmless products.

2.3.2.3 Sulfide-Producing Anaerobic Zone

After all the oxygen has been removed from the aerobic zone (if present,) bacteria in the sulfide-producing anaerobic zone use combined sources of oxygen such as nitrate (NO_3^-), nitrite (NO_2^-), sulfite ($SO_3^=$) and sulfate ($SO_4^=$) to carry out their life processes. When these bacteria use nitrate and nitrite they produce nitrogen gas (N_2) which has no odor, and when they use sulfite and sulfate they produce sulfide ($S^=$) and hydrogen sulfide gas.

When compounds containing combined oxygen (such as nitrate and sulfate) are converted to compounds without oxygen (such as nitrogen gas or hydrogen sulfide) they are said to be "reduced" because they lose oxygen. When sulfide is converted back to sulfate in the aerobic zone, it is said to be "oxidized" because it gains oxygen. Some of the bacteria in this zone can use nitrate and are called nitrate-reducing bacteria, or "nitrate reducers," whereas some can use sulfate and are called "sulfate-reducing bacteria" (SRB). In general, bacteria can extract more energy from nitrate than from sulfate. Therefore, when nitrate is present in the wastewater, nitrate-reducing bacteria in this zone become active and dominate the slime layer, even consuming the sulfide being produced by their neighbors. Since nitrogen gas is produced, this process generates no odor. However, nitrate and nitrite are rarely present naturally in domestic wastewater and so must be artificially added (see Chapter 7). However, sulfate is almost universally found in drinking water, and concentrations are always high enough in domestic wastewater to produce sulfide.

It is important to note that even if an aerobic zone or an artificial nitrate dominated zone is present on the outer surface of the slime layer, sulfide production still continues below. Sulfate can migrate through the aerobic and nitrate dominated zones completely unaffected and enter the sulfide producing zone where it is converted to sulfide. However, any sulfide produced under these circumstances would be quickly oxidized back to sulfate as it diffuses back out of the slime layer resulting in zero or near zero release of sulfide. It will appear as if the addition of oxygen or nitrate "stopped" the production of sulfide, which is technically incorrect. The sulfide is still produced but is actually being reoxidized to sulfate.

2.3.2.4 Anaerobic Zone

This zone does not directly produce sulfide and has historically been called "inert" for this reason. However, it plays an important role in sulfide production and is not actually inert. The food (BOD_5) that penetrates all the way to this zone is converted anaerobically to fermentation products. Fermentation products are small organic compounds left over after anaerobic bacteria are finished. Strict anaerobic bacteria that live here can only break down organic compounds to the two or three carbon atom stage even though there is energy left in them. Only aerobic bacteria or bacteria in the sulfide-producing zone can break them down further to retrieve energy. As these fermentation products migrate back up through the sulfide-producing zone, bacteria use them to convert sulfate to sulfide. Although this is not the only source of food for the SRBs, it can be a significant contribution to sulfide production.

From the discussion above, we can see that the rate at which sulfide is produced by the slime layer depends on a variety of environmental conditions including the concentration of organic food source (BOD), dissolved oxygen concentration, presence of nitrate, sulfate concentration, and the submerged surface area of the pipe as well as other factors.

2.3.2.5 Preventing Sulfide Generation

It follows from the above discussion that sulfide generation in wastewater collection systems could be reduced or minimized by some or all of the following:

1. Minimize or reduce wastewater BOD concentration, particularly soluble BOD sources such as sugars, soluble lipids and readily hydrolysable proteins.
2. Minimize or reduce the sulfate concentration of wastewater if possible through industrial discharge controls.
3. Minimize or reduce elevated temperature discharges.
4. Minimize or reduce hydraulic detention time in the collection system.
5. Minimize the use of force mains and consider twinning force mains for future capacity as opposed to a single larger main.
6. Minimize or reduce sediment and debris in sewers through regular assessment and cleaning as required.

7. Hydraulically scour and remove a portion of the slime layer from force mains by periodically allowing the wet well to fill and turn on all pumps. High velocity water will remove some of the bacteria responsible for sulfide generation. Required on a weekly routine or more often as needed.

Regardless of the rate of sulfide generation, when oxygen and nitrate are not present, there is no mechanism to reoxidize sulfide on its way out of the sulfide-producing zone. Under these conditions, sulfide is released back into the wastewater stream, where it quickly comes into chemical equilibrium between four different forms of sulfide: the sulfide ion ($S^{=}$), the bisulfide (or hydrosulfide) ion (HS^-), aqueous hydrogen sulfide ($H_2S_{(aq)}$), and hydrogen sulfide gas ($H_2S_{(g)}$). This chemical equilibrium and the properties of each sulfide chemical in it are critical to understanding the control of sulfide in wastewater collection systems. Each of the sulfide chemical species is briefly described below.

Sulfide ion ($S^{=}$). The sulfide ion carries a double negative charge, indicating that it reacts primarily by giving up or sharing two electrons from the outer shell. It is a colorless ion in solution and cannot leave wastewater in this form. It does not contribute to odors in the ionic form.

Bisulfide ion (HS^-). The bisulfide (or hydrosulfide) ion carries a single negative charge. This is because one of the negative charges of the sulfide ion is taken up by a positively charged hydrogen ion. The hydrogen ion (H^+) was taken away from a water molecule (H_2O), leaving one hydroxide ion (OH^-) behind. The hydroxide ion increases the alkalinity (pH) of the wastewater and serves an important function to be discussed later. The bisulfide ion is a colorless, odorless ion that can exist only in solution. It also does not contribute to odors in this form.

Hydrogen sulfide (aqueous) ($H_2S(aq)$). Hydrogen sulfide can exist as a gas dissolved in water. The polar nature of the hydrogen sulfide molecule makes it moderately soluble in water. In the dissolved (aqueous) form, hydrogen sulfide does not cause odor; however, this is the only sulfide compound that can leave water to exist as a free gas. The rate at which hydrogen sulfide leaves water is governed by Henry's law, how turbulent the wastewater is, and the pH of the solution.

Hydrogen sulfide (gaseous) ($H_2S(g)$). Once hydrogen sulfide leaves water and becomes a gas, it can cause odor and corrosion. Hydrogen sulfide gas is a colorless but extremely odorous gas that can be detected by the human sense of smell in concentrations as low as 0.0005 parts per million (ppm), although 0.002 ppm is probably more near the average threshold. It is also very hazardous to humans in high concentrations and can cause a number of health-related problems and even death. In concentrations as low as 10 ppm it can cause nausea, headache, and eye irritation. Above 100 ppm it can cause serious breathing problems and loss of the sense of smell along with burning of the eyes and respiratory tract. Above 300 ppm death can occur within a few minutes. Because it can deaden the human sense of smell in concentrations above 100 ppm, it is particularly dangerous, making victims think that it is no longer present. Hydrogen sulfide is the cause of death in numerous wastewater accidents every year. A generally accepted hydrogen sulfide odor and toxicity spectrum is presented in Figure 2-3.

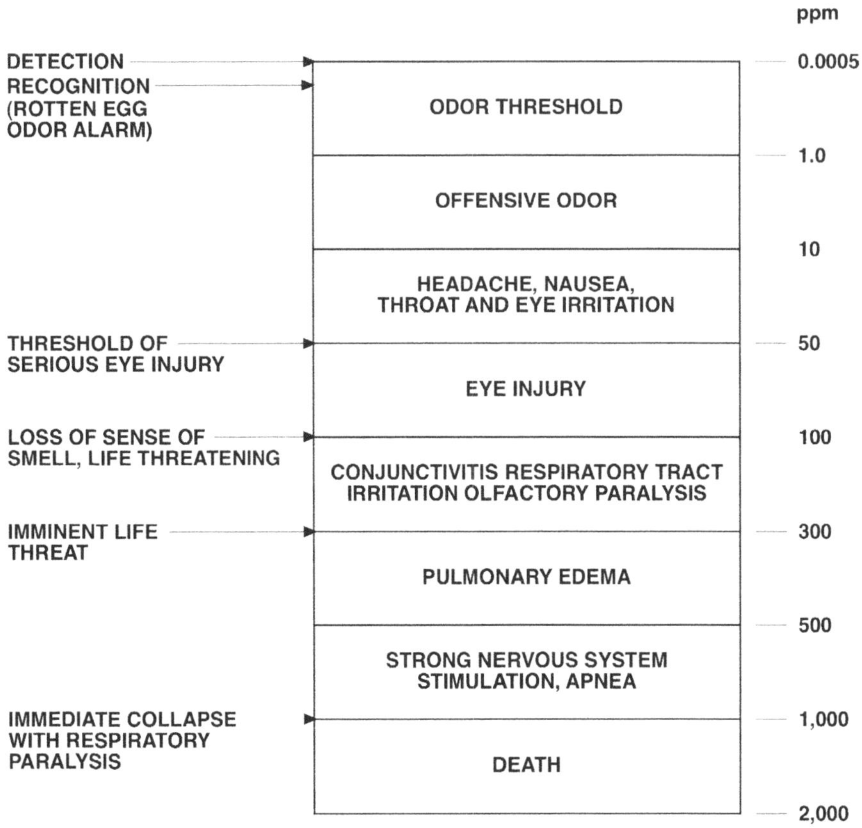

Figure 2-3. Generally Accepted Toxicity Spectrum for Hydrogen Sulfide Gas.

Due to the continuous production of sulfide by the slime layer, hydrogen sulfide gas rarely, if ever, re-enters the liquid phase. Sulfide continuously produced by the slime layer replaces that which is lost to the atmosphere as hydrogen sulfide gas in the collection system. In addition, once the hydrogen sulfide gas is released it usually disperses throughout the sewer environment and never reaches a high enough concentration to be forced back into solution.

The four sulfide chemical compounds are related according to the following equilibrium:

$$H_2S(g) \rightleftharpoons H_2S(aq) \rightleftharpoons HS^- \rightleftharpoons S^= \qquad (2.3.2\text{-}1)$$

As indicated by the equilibrium equations, once dissolved hydrogen sulfide is released into the gas phase, the bisulfide ion is immediately transformed into more aqueous hydrogen sulfide to replace that which was lost. Concurrently, sulfide ion is transformed into bisulfide to replace that lost to aqueous hydrogen sulfide. The equilibrium shift is very rapid and sensitive to the pH of the wastewater.

2.3.3 Relationship Between Turbulence and Sulfide Stripping

Turbulence is a critical parameter to consider in preventing hydrogen sulfide gas release from wastewater. The effects of sulfide odor and corrosion are increased significantly at points of turbulence. Turbulence, splashing and hydraulic outfalls generate water droplets which dramatically increase the surface area of the liquid. Any action that serves to increase the surface area of the liquid also increases the rate of H_2S gas transfer from the liquid to the gas phase.

Unnecessary turbulence should be avoided in collection systems whenever possible. If turbulence is unavoidable due to drops, steep slopes or outfalls, the effects of odor and corrosion should be mitigated during design before they cause a problem.

The quantitative relationship between two of the most important sulfide species (HS^- and $H_2S_{(aq)}$) is controlled by the pH of the wastewater as shown in Figure 2-4. This figure indicates that the proportional relationship between these two sulfide species is approximately 50/50 at a pH of 7.1. This means that at the pH of normal wastewater, half of all sulfide exists as the bisulfide ion and the other half exists as aqueous hydrogen sulfide ($H_2S_{(g)}$—a dissolved gas). Therefore, half of all sulfide in normal domestic wastewater is in a form that can leave the liquid phase to exist as hydrogen sulfide gas. If the pH of the wastewater is lowered, more of the sulfide exists as hydrogen sulfide gas.

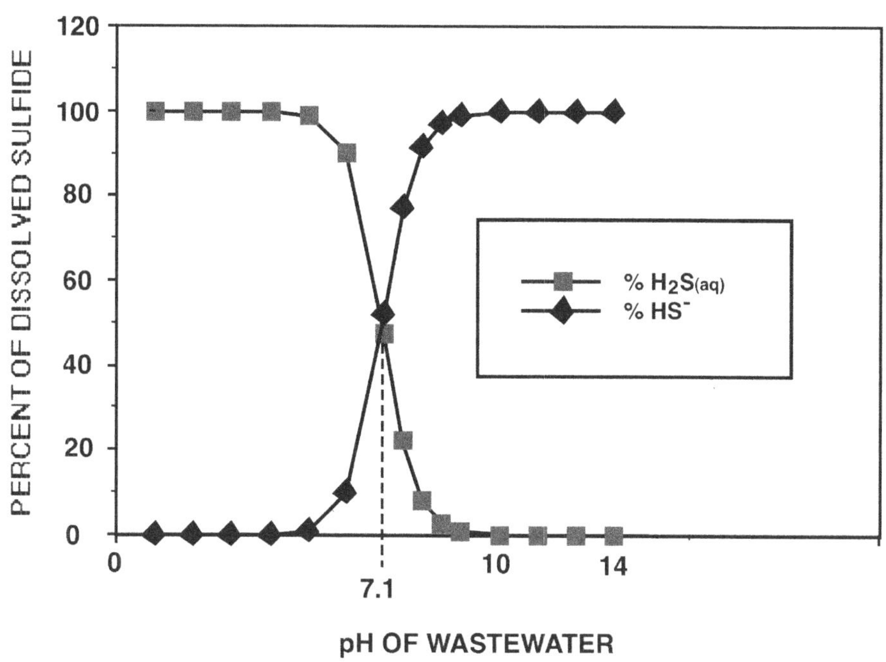

Figure 2-4. Effect of pH on Dissolved Sulfide Form.

The sulfide equilibrium is extremely delicate and can shift rapidly with changes in temperature, barometric pressure, agitation, and particularly pH. As illustrated in Figure 2-4, the relationship curves between these two sulfide species is very steep near neutral pH. This means that a small shift in pH can translate into a large difference in hydrogen sulfide gas release. This fact has both negative and positive consequences. Low-pH discharges can cause increased localized release of hydrogen sulfide gas, which in turn causes increased odors and corrosion. On the other hand, artificially increasing the pH of wastewater to prevent hydrogen sulfide release is sometimes a viable sulfide control alternative. In one case, shifting the wastewater pH from 6.8 to 8.2 reduced hydrogen sulfide gas concentrations over 90% at a wastewater treatment plant headworks.

2.3.3.1 Preventing Sulfide Release

It follows from the above discussion that sulfide release from wastewater collection systems could be reduced or minimized by some or all of the following actions:

1. Minimizing or reducing wastewater turbulence in manholes, wet well entrances, and sewers through careful design and hydraulic modification.
2. Using contained-drop pipes for drop structures to minimize droplet formation and offgassing.
3. Inspecting and cleaning sewers where debris accumulation could occur or is suspected to have occurred. Debris in sewers can cause turbulence as well as increased sulfide generation.
4. Considering raising industrial discharge pH limits to allow higher wastewater pH and reducing sulfide-stripping potential.

2.4 Odorous Compounds in Wastewater

There are many odorous substances found in wastewater collection systems in addition to hydrogen sulfide gas. Hydrogen sulfide gas is inorganic, meaning it does not contain carbon. Hydrogen sulfide and ammonia are two examples of inorganic odor compounds found in wastewater collection systems. However, there are many more organic odorous compounds that can be produced in wastewater. All organic odor compounds contain carbon in one form or another. In addition, almost all of these organic odor compounds contain sulfur in some form. Some of the more common odor-causing substances, along with their chemical formula and odor threshold concentration, are listed in Table 2-1. All of these compounds can be produced naturally from domestic wastewater and sludge. Hydrogen sulfide, ammonia, and sulfur dioxide, which are inorganic, are included in Table 2-1 for comparison of their odor characteristics.

Table 2-1. Odorous Compounds in Wastewater.

Name	Formula	Characteristic Odor	Detection Threshold (ppm)
Hydrogen sulfide	H_2S	Rotten eggs	0.0005
Ammonia	NH_3	Irritating, pungent	17
Skatole	C_9H_9N	Fecal, nauseating	0.001
Indole	$C_6H_4(CH)_2NH$	Fecal, nauseating	0.0001
Methylamine	CH_3NH_2	Putrid, fishy	4.7
Allyl mercaptan	$CH_2{=}CHCH_2SH$	Disagreeable, garlic	0.0001
Amyl mercaptan	$CH_3(CH_2)_4SH$	Unpleasant, putrid	0.0003
Benzyl mercaptan	$C_6H_5CH_2SH$	Unpleasant, strong	0.0002
Ethyl mercaptan	C_2H_5SH	Decayed cabbage	0.0003
Dimethyl sulfide	$(CH_3)_2S$	Decayed cabbage	0.001
Trimethylamine	$(CH_3)_3N$	Pungent, fishy	0.0004
Sulfur dioxide	SO_2	Pungent, irritating	2.7
Methyl mercaptan	CH_3SH	Decayed cabbage	0.0005
Thiocresol	$CH_3C_6H_4SH$	Skunk, irritating	0.0001
Thiobismethane	CH_3SCH_3	Rotting meat	0.0011

Note: Different threshold values are reported in the literature, particularly for hydrogen sulfide and ammonia. Odor thresholds are included here to give the reader a general idea of human sensitivity and odor concentrations which could potentially cause complaints. Ammonia for example is reported in ranges from 2.8 ppm to 17 ppm.

Because Henry's law also applies to organic compounds, the potential for stripping of these compounds is increased at points of turbulence within collection systems and treatment processes.

Although numerous odor compounds are produced in wastewater and released into sewer systems, the most common compound associated with odor complaints is hydrogen sulfide. This is due partly to its universal presence in wastewater collection systems and partly to the human ability to detect the gas at extremely low concentrations. If air containing hydrogen sulfide gas and organic odor compounds escapes a collection system, hydrogen sulfide gas will typically dominate the odor as it disperses into the environment. However, the escaping air will also contain organic odor compounds that can cause complaints when hydrogen sulfide is absent.

Since human reactions to odor chemicals in the air are subjective and vary from person to person, odors cannot be easily measured or quantified with a straightforward analytical approach. The problem then becomes, How do I measure the impact of odor from my facilities? It is possible to sample air directly for some of the odor-causing compounds. Many regulatory authorities establish odor regulations and compliance criteria based upon the measured concentration of certain odor compounds (most often hydrogen sulfide gas). Many sampling methods and analytical techniques have been developed to measure specific odor compounds released from wastewater (see Chapter 4.0).

Odors also have different characters, with varying degrees of offensiveness, called the Hedonic tone of the odor. Hedonic tone is a very important parameter for odors and explains why some odors can produce serious complaints at very low concentrations whereas others can be tolerated by humans well beyond their odor threshold concentrations. Knowing which odors are more likely to produce complaints is important when considering which odor sources to control with limited budgets. To gain insight into the nature of odors and what makes them different from each other, it is necessary to briefly look at how the human brain recognizes and reacts to odors.

2.4.1 Odors 101

As discussed above, odor compounds are simply volatile chemical compounds in the air that have an unpleasant character. Volatile chemicals with a pleasant odor character are called "perfumes" instead of odor compounds. Whether pleasant or unpleasant, a volatile chemical in the air that can be detected with the human sense of smell is called an "odorant." It is important to note that at any given time there could be hundreds if not thousands of volatile chemicals (odorants) in the air around you, depending upon where you are. Only a very few of these chemicals are recognized by the human brain as unpleasant or offensive. How are these volatile chemical compounds recognized, and why are only some of them offensive to humans?

The human sense of smell is a very sensitive, broad-spectrum chemical analyzer with a direct link to the brain. The effect that volatile chemicals (both pleasant and unpleasant) can have on the human brain is controlled by the sensitivity of the sense of smell, which varies from person to person. However, there are certain traits of the human sense of smell (the olfactory sense) that are common to all individuals. Understanding a little about how the human olfactory system and brain work together to recognize odors will help us better understand which compounds might require priority attention and how best to control them.

Imagine that you are walking down a path in a quiet forest on a warm, sunny spring day. The air is filled with hundreds of odorants of many types and forms. There are volatile chemicals being released by the growing leaves, bark, and flowers on the trees and the decaying leaves from the previous fall; flowers are releasing pollen and insect-attracting chemicals, molds are releasing spores, insects and animals are releasing pheromones, and fungi on the rotting logs you occasionally step over are releasing many volatile decay products. However, your olfactory sense is telling your brain that the air smells fresh with earthy aromas and has a generally pleasant character, so you are at ease with your environment. No unusual, dangerous, or extraordinary odors have been detected so your brain does not bother to bring them to your conscious attention.

Now imagine that you come across the rotting carcass of an animal along the path. Before you see it, your olfactory sense picks up the chemical, recognizes it as being associated with a potentially unhealthy environment and sends a warning to your brain. Your brain immediately brings the odor to your conscious attention, prompting you to look for the source of the odor in order to avoid contact with it. Your brain is acting solely on the odor input to protect you from what it perceives as a potentially dangerous situation. If you can identify the source and avoid contact with it, then your olfactory senses and brain have done their job and you can return to enjoyment of the day. However, if the source of the odor cannot be identified or avoided (one or both are often the case with wastewater odor complaints) your brain remains in a state of alert and you are not at ease with your environment.

What is the difference between these two situations? Why does your brain tolerate hundreds of volatile chemicals and conclude that you are in a pleasant environment but cause significant distress at the presence of one or two particular chemicals?

The volatile chemicals you recognized as pleasant, earthy fragrances were generated by aerobic processes and the unpleasant decaying odor was generated by anaerobic bacteria. We live in an aerobic environment where oxygen is plentiful. We are exposed to the products of aerobic life on a daily basis, we eat mostly aerobically produced foods and we are familiar with them. Our bodies have very little to fear from aerobic environments or the products of aerobic systems. Therefore, aerobic decay products are similarly not harmful to us and humans have not developed olfactory warning systems for most aerobic compounds.

However, some chemical compounds are automatically recognized by the human olfactory system and brain as dangerous or threatening. These compounds are produced by anaerobic processes in environments which could be potentially harmful to us. Our bodies do not tolerate anaerobic conditions well and we cannot eat most anaerobic things without becoming ill. Our feces as well as the feces of most animals is anaerobic and unhealthy, prompting us to recognize these odors and avoid contact. Many gasses produced by anaerobic processes (swamp gas) are poisonous to humans. Once recognized by the brain, anaerobic odors produce an immediate "flight" response which does not require conscious thought. The detection of these compounds is "pre-wired" into the human brain to protect us from potential environmental dangers.

One of these compounds is hydrogen sulfide gas. Because hydrogen sulfide can be deadly to humans, we are automatically programmed to detect and be repelled by the characteristic "rotten egg" odor. This protects humans against possible health impact and death.

Many *organic* (carbon-containing) odor compounds are produced by anaerobic decay and are also characterized as being "rotten" or "decaying." This response also protects us against biological hazards by warning us not to eat or get near certain things. Humans rely on the link between their sense of smell and taste to warn them of contaminated, rotten or spoiled foods which could cause them harm. In the rare case that an individual were not able to distinguish the difference between fresh and rotting food, they would likely not live long enough to pass this defective trait to offspring.

The olfactory detection of an odor compound that our brain recognizes as a potential threat to our health causes stress at the primal (subconscious) level of our brain. That means that olfactory signals bypass the logic and reason processing centers of our brain to elicit the "flight" response. There are many other factors which affect the severity of human responses to environmental odors. The concentration of the odor chemical in the air, its duration, and its frequency of occurrence (which are not discussed here) all affect the human reaction to environmental odors. This brief discussion only serves as an introduction to explain why some odor compounds cause a negative human perception and others do not. Knowledge of how odors impact humans can help identify odor sources deserving priority attention from those of lower impact and priority. This knowledge also helps us select more appropriate and cost effective odor treatment systems and technologies.

2.4.2 Sulfuric Acid Production

Hydrogen sulfide gas is not only the most common cause of odor complaints from wastewater collection systems; it is also responsible for 99% of all sewer corrosion. There is another type of aerobic bacteria that commonly colonizes pipe crowns, walls, and other surfaces above the waterline in wastewater pipes and structures. These bacteria have the ability to consume hydrogen sulfide gas, oxidize it for use as an energy source and excrete sulfuric acid. This process can take place only where there is an adequate supply of hydrogen sulfide gas (>2.0 ppm), high relative humidity, carbon dioxide gas, and atmospheric oxygen. These conditions exist in the majority of wastewater collection systems for at least some portion of the year. The pH of surfaces exposed to severe hydrogen sulfide environments (>50 ppm in air) have been measured as low as 0.5, which is approximately equivalent to a 7% sulfuric acid solution.

The simplified and balanced equation for the biological metabolic process which converts hydrogen sulfide to sulfuric acid is presented below:

Thiobacillus bacteria

$$H_2S(g) + 2O_2 \rightarrow H_2SO_4 \qquad (2.4.2\text{-}1)$$

hydrogen sulfide, atmospheric, sulfuric

gas, oxygen, acid

2.4.3 Concrete Corrosion

The effect of sulfuric acid on concrete surfaces exposed to the sewer environment can be devastating. Sections of collection interceptors and entire pump stations have collapsed due to loss of structural stability from sulfide-related corrosion. We have all seen the photographs of cars being swallowed by streets opened by the loss of a section of large-diameter sewer (and also water main breaks). The process of concrete corrosion, however, is a step-by-step process that can sometimes give misleading impressions. The following briefly describes the biochemical process of concrete corrosion in the presence of hydrogen sulfide in a sewer atmosphere.

Freshly placed concrete has a pH of approximately 12 or 13, depending upon the amount of cement in the mix design. This high pH is the result of the formation of calcium hydroxide [$Ca(OH)_2$] as a byproduct of the hydration of cement. Calcium hydroxide is a very caustic crystalline compound which can occupy as much as 25% of the volume of concrete. Any surface with a pH of 12 or 13 will not allow the growth of any bacteria, because these very high pH values will dissolve the protein in the cell walls. However, the pH of the concrete is lowered over time by the effect of carbon dioxide (CO_2) and hydrogen sulfide gas (H_2S). These gases are both known as "acid" gases because they form relatively weak acid solutions when dissolved in water: CO_2 produces carbonic acid, and H_2S produces thiosulfuric and polythionic acids. These gases dissolve into the water on the moist, high-pH concrete surfaces above the sewage flow and react with the calcium hydroxide to slowly reduce the pH of the surface. Eventually the surface pH is reduced to a level which can support the growth of bacteria (pH 9 to 9.5).

The time it takes to reduce the initial high pH of fresh concrete is a function of the concentration of carbon dioxide and hydrogen sulfide in the sewer atmosphere. It can sometimes take years to lower the pH of concrete from 13 to 9; however, in some severe situations it can be accomplished in a few months.

Once the pH of the concrete is reduced to around 9, biological colonization can occur. Over 60 different species of bacteria are known to regularly colonize wastewater pipelines and structures above the water line. Most species of bacteria in the genus *Thiobacillus* have the unique ability to convert hydrogen sulfide gas to sulfuric acid in the presence of oxygen. Because each species of bacteria can survive only under a specific set of environmental conditions, the particular species inhabiting the colonies changes with time. Since the production of sulfuric acid from hydrogen sulfide is an aerobic biological process, it can occur only on surfaces exposed to atmospheric oxygen above the waterline in sewers.

As a simplified example, one species of *Thiobacillus* grows well only on surfaces with a pH between 9 and 6.5. However, when the sulfuric acid waste product they excrete decreases the pH of the surface to below 6.5, they die off and another species that can withstand lower pH takes up residence. The succeeding species grows well on surfaces with a pH between 6.5 and 4. When the acid produced by this species drops the pH below 4, a new species takes over again. The process of successive colonization can continue until species take over which can survive under extremely low pH conditions. One such species is *Thiobacillus thiooxidans*. This organism has been demonstrated to grow well in the laboratory while exposed to a 7% solution of sulfuric acid. This is equivalent to a pH of approximately 0.5.

Sulfuric acid attacks the matrix of the concrete, which is commonly composed of calcium silicate hydrate gel (CSHG), calcium carbonate from aggregates (when present), and unreacted calcium hydroxide. Although the reaction products are complex and result in the formation of many different compounds and minerals, the biogenic corrosion process can be generally illustrated by the following three reactions:

$$H_2SO_4 + CaSi\ (CSHG) \rightarrow CaSO_4 + Si + 2H^+ \qquad (2.4.3\text{-}1)$$

$$H_2SO_4 + CaCO_3 \rightarrow CaSO_4 + H_2CO_3 \qquad (2.4.3\text{-}2)$$

$$H_2SO_4 + Ca(OH)_2 \rightarrow CaSO_4 + 2\ H_2O \qquad (2.4.3\text{-}3)$$

The primary product of concrete decomposition by sulfuric acid is calcium sulfate ($CaSO_4$), more commonly known by its mineral name, gypsum. From experience with this material in its more common form of drywall board, we know that it does not provide much structural support, especially when wet. It is usually present in sewers and structures as a pasty white mass on concrete surfaces above the water line. In areas where diurnal or other high flows intermittently scour the walls above the water line, concrete loss can be particularly fast.

It is generally believed by most investigators that the surface coating of gypsum paste somewhat protects underlying sound concrete by providing a buffer zone through which freshly produced sulfuric acid must penetrate. Because *Thiobacillus* bacteria are aerobic, they require free atmospheric oxygen to survive. Therefore, they can live only on the thin outer covering of any surface. This means that acid produced on the surface must migrate through any existing gypsum paste to reach sound concrete. By washing off the "protective" coating of gypsum with high diurnal or wet weather flows, fresh surfaces are repeatedly exposed to acid attack, which accelerates the corrosion process. Sewer-cleaning practices and equipment should be checked to determine the degree of gypsum layer loss during cleaning and the effects of such cleaning on the ultimate pipe life. Frequent high-pressure cleaning of concrete sewers under the influence of hydrogen sulfide could accelerate the corrosion process.

2.4.4 Metal Corrosion

Concrete is not the only material which can fall prey to the corrosive effects of hydrogen sulfide gas and sulfuric acid. Most metals, including many stainless steel alloys, can also be attacked and destroyed by exposure to sulfuric acid. Metals, however, are offered two routes of attack: acidic decomposition by exposure to sulfuric acid produced by *Thiobacillus* bacteria and direct chemical attack. Most common metals can exist as divalent cations, meaning that they carry two positive charges and react primarily by gaining two electrons in their outer shell (M^{++}). The sulfide component of hydrogen sulfide gas supplies these two electrons, resulting in a metal sulfide and two free hydrogen ions.

$$M^{++} + H_2S \rightarrow MS + 2H^{+} \tag{2.4.4-1}$$

The metal has a much stronger affinity for the sulfide than for the hydrogen, resulting in the formation of a stable sulfide-metal product and the release of two free hydrogen ions. In a moist or wet environment the hydrogen ions lower the pH of the water and increase the rate of corrosion. By reacting with sulfide, the metal is converted from the strong metal-metal bonding into a much weaker metal-sulfide product, and at the same time is exposed to the acidic effect of the free hydrogen ions. Most wastewater professionals have witnessed the corroded rungs of steel manhole steps, corroded and weakened manhole covers and rings, and have seen brass and copper fittings turned dark bluish-black, the color of nickel and copper sulfide.

2.5 Trends in Sulfide Production

Operators know that the production of hydrogen sulfide in wastewater collection systems has been increasing over the past 15 or so years. There are now significant scientific data to confirm that sulfide generation, odor, and corrosion are increasing in today's wastewater collection systems. A summary of some of the seemingly unrelated reasons for this increase is presented below, along with implications for collection system operators.

2.5.1 Clean Water Act

The Federal Water Pollution Control Act Amendments of 1972 (PL 92-500), as supplemented by the Clean Water Act (CWA) of 1977, mandated that the nation's water supplies be protected through a variety of mechanisms. One such mechanism was to require cleaner wastewater discharges from our publicly owned treatment works. One concern with treated wastewater discharges was the bioaccumulation of heavy metals in the environment. Certain metals, such as lead, copper, mercury, chromium, and zinc, were shown to have toxic effects to animals and plants in high concentration. Some of these metals passed through wastewater treatment plants and entered the receiving stream where they could be bioaccumulated to dangerous concentrations. As a result, the U.S. EPA required municipalities to implement industrial pretreatment programs where heavy metals and other contaminants from industrial sources were identified. Before 1980, a common source of heavy metals in wastewater was industrial discharges from steel mills, electroplating operations, photo-finishing, and electronics and computer chip manufacturing. In 1995, all such operations were required to begin pretreating their wastewater to remove metals prior to discharging the wastewater to a publicly owned treatment work.

Sulfide in the dissolved forms (bisulfide ion and sulfide ion) has a strong affinity for metals. In fact, one of the most common methods used to remove sulfide from wastewater is to add metal salts (usually ferrous iron compounds) to combine with sulfide and precipitate metal sulfide, which is very insoluble. The substitution of iron for other heavy metals does not violate the industrial pretreatment standards since iron is essentially not regulated and is abundant in nature. Most iron sulfide is removed at the treatment plant as sludge, but iron is typically not a permitted discharge constituent. Sulfide will combine with almost all metals; therefore, removing a source of metals in wastewater also removes a potential sulfide-removal mechanism.

It has also been noted that the targeted heavy metals exhibited a toxic effect on the slime layer bacteria that produce sulfide. Toxicity or inhibition of the sulfate-reducing bacteria in the slime layer naturally reduced their health and activity, which in turn reduced sulfide production. Removing the source of the toxicity allowed the slime layer bacteria to flourish and produce even more sulfide. In 1989, congress commissioned a study to assess the impact of the industrial pretreatment program on sulfide odor and corrosion. The study concluded that both of the above mechanisms were responsible for a general increase in sulfide concentrations in domestic wastewater. Figure 2-5 represents data taken at the Hyperion Wastewater Treatment Plant, in Los Angeles, California.

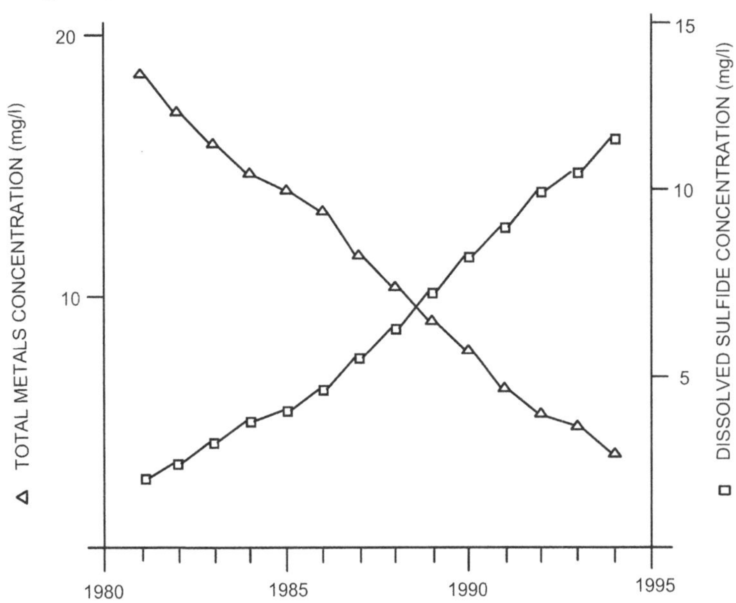

Figure 2-5. Metals-versus-Sulfide Relationship.

Over a period of several years, influent wastewater was analyzed for several constituents, including sulfide and total metals concentrations. As the data dramatically indicate, influent sulfide to the plant increased at the same time total metals concentrations decreased. This trend began in 1980 and continues to this day; it can therefore be concluded that industrial pretreatment has resulted in an increase in sulfide concentrations in wastewater and thus in increased odor and corrosion.

2.5.2 Water Conservation Practices

Over the last 15 years, utilities and water purveyors have recognized the benefit of promoting water conservation practices. Prompted by water shortages around the country, water conservation practices were shown to not only preserve a precious resource but also save utilities money by delaying planned expenditures for plant upgrades, system capacity increases, and storage facilities. Although cost-efficient and practical for the water industry, water conservation has caused an increase in wastewater system dissolved sulfide concentrations.

Reduced wastewater flows from water conservation practices cause reduced flow velocities and longer residence times, or the time wastewater spends in the collection system. This allows more time for the reduction of sulfate and therefore higher dissolved sulfide concentrations. It also increases the general septicity of the wastewater. Since less dilution water is entering the wastewater collection system, the strength of the wastewater in terms of BOD_5 increases. This increases the biological activity of the slime layer and causes the faster consumption of dissolved oxygen. Increasing septicity causes the production of organic acids that drop the pH of the wastewater. Since a small shift downward in the pH of the wastewater can cause a dramatic increase in hydrogen sulfide gas release (see Figure 2-4), the anaerobic drop in pH can also increase the release of odor-causing hydrogen sulfide. Although water conservation has increased the dissolved sulfide concentration of wastewater, there is little evidence to suggest this alone increases total sulfide mass. Increased sulfate or BOD_5 would be required to cause this effect. However, the Henry's law coefficient governing the release of hydrogen sulfide gas from wastewater is highly dependent upon sulfide concentration in the liquid phase. Therefore, any increase in the dissolved sulfide concentration of wastewater will increase the likelihood of hydrogen sulfide gas releases along with the associated odor and corrosion.

2.5.3 Regionalization of Wastewater Treatment Facilities

Over the past several years, wastewater agencies have been regionalizing wastewater treatment facilities, building larger plants or upgrading others while closing smaller, less-efficient plants. This action follows from the realization that a few larger plants are less costly to operate than many smaller ones. Also, improving many small plants to keep up with tightening regulatory requirements is much more expensive than upgrading one or two larger plants. The usual method of regionalizing wastewater treatment is to build a pumping station at the site of the small treatment plant, pump the wastewater to a larger or newer plant and demolish the small plant. This requires the use of pumping stations and force mains that have a 360° slime layer, have an increased detention time, do not allow reaeration, and are excessive producers of sulfide. The sulfide produced in the force mains is released when it is exposed to the atmosphere.

In addition, municipalities are sewering larger and more extensive areas of their districts in an effort to minimize the use of septic tanks. This is an effort to protect local groundwater supplies. Usually, the wastewater piping is oversized in anticipation of larger future capacities; this causes low velocities, long residence times, and excessive sulfide production.

2.5.4 Culture and Environmental Awareness

One of the reasons for increased odor problems is a general increasing awareness of odor and a lowering of the intolerance for it. Many years ago in the United States, and still in some other countries, odor was a given fact of life and was considered an acceptable price to pay for progress and prosperity. This was before an increasing general awareness of the environment in the late 1960s and 1970s. As people became more aware of the environment and its workings, odor became a less and less accepted part of everyday life. It was perceived that if offensive odors are detectable, there must be something environmentally wrong. And when that odor had something to do with wastewater collection or treatment, then the environment, or at least people's health, must surely be threatened.

2.5.5 Urbanization

As population grows, land becomes more intensively used. This is particularly true of the areas near wastewater treatment plants and other wastewater facilities. Utilities often go to great lengths to locate treatment plants and other facilities as far away as possible from developed areas. Unfortunately, developers recognize the benefit of readily available roads, power and water lines installed to support the plant. Add to this the ready availability of sewer access and the land leading to and around the treatment plant is soon filled with new homes. This scenario happens all too often. Intensifying land use near wastewater facilities does not increase sulfide production, but it does make life difficult for wastewater professionals, increases capital and O&M budget expenditures for odor control measures, and can be a significant public relations problem.

The fact that wastewater no longer contains the natural control afforded by metals in industrial discharges, and that stronger wastewater is now traveling farther than ever to get to its final treatment destination is sufficient reason to be concerned with odors from wastewater systems. Add to this the fact that people are living closer to wastewater facilities and are no longer willing to put up with odors, and it is no small wonder that we are now faced with even greater odor and corrosion challenges for the future.

2.6 The Sulfur Connection

We have seen above how odors and corrosion in wastewater collection systems cause complaints from citizens and damage our wastewater infrastructure. We have also found that the principal chemical compound implicated in both cases is hydrogen sulfide gas. We know that it is produced in the collection system through the interaction of bacteria and the waste products in the sewer. We know that a great amount of money is spent each year in attempts to control this compound. How can a simple, plentiful compound like sulfur cause such damage? We must first investigate the "sulfur connection" to see how hydrogen sulfide is formed. This will allow us to identify our options for controlling odor and corrosion in collection systems.

2.6.1 In Nature

Sulfur is one of the most common elements on earth. It is abundantly present in rocks, minerals, soil, and water. Because sulfur is soluble in most of its forms, groundwater and surface water can contain considerable quantities of sulfur, depending upon local geology. In addition, seawater contains high concentrations of sulfur as sulfate, as well as sulfide and sulfite. The ocean is also a huge repository for organic sulfur accumulated from aquatic life processes such as seaweed and plankton deposits. Undersea volcanic activity also spews thousands of tons of sulfur into the oceanic abyss every hour. Sulfur in one or more if its forms is intimately connected with life on Earth.

As an element in nature, sulfur can exist in a wide variety of chemical forms. Because of the configuration of the outer electron shell for sulfur, it is capable of giving up, receiving, or sharing electrons in a variety of combinations. It is therefore able to combine with many other elements and compounds and has become widely distributed in nature. As a result of its wide natural distribution, sulfur has become an integral part of our ecosystem and biosphere. Most living things require sulfur in some form to carry out their life processes. It serves important biochemical functions in many plants and animals. It is an essential component of many proteins and a critical component of blood plasma.

Some species of bacteria have evolved to use sulfur as the primary component in their life cycle. Some bacteria require certain sulfur compounds to be present before their metabolism can function, while others can use only certain specific sulfur compounds as terminal electron acceptors. In extreme environments (such as undersea volcanic fumaroles), entire ecosystems have even evolved that use inorganic sulfur as a total energy source, without reliance on energy from the sun. Needless to say, the life processes of thousands of different types of bacteria are centered on sulfur in one or more of its various forms.

Almost any good textbook on environmental chemistry will contain a figure similar to Figure 2-6, which depicts the sulfur cycle and its many transformations in nature.

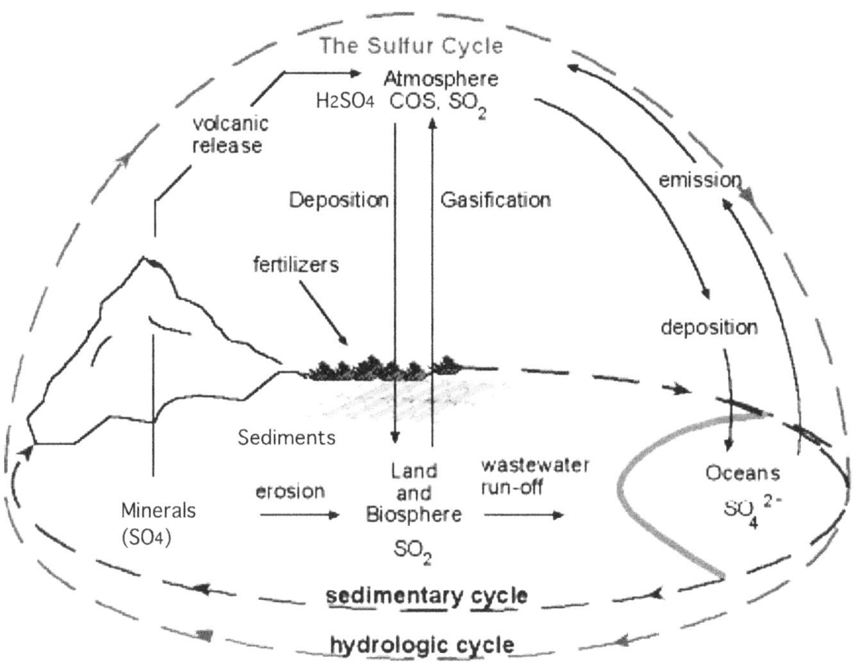

Figure 2-6. The Sulfur Cycle.

The sulfur cycle is an important tool for learning about the overall environmental, chemical, and biological significance of sulfur, but it does not readily explain the small but important environmental niche that exists for sulfur in municipal wastewater collection systems.

2.6.2 In the Sewer

Civilization has created a very specialized environmental niche we call "the sewer." This niche consists of a complex biochemical environment that is populated by many different life forms. It also exhibits a specific and definable, semiclosed biological and chemical cycle of sulfur that can be identified and observed. The cycle is based upon the biological and chemical transformation of sulfate into sulfide and sulfide into a variety of ionic and gaseous forms, ending in the biological transformation of hydrogen sulfide gas to sulfuric acid and then neutralization of the acid back to sulfate. This is the cycle that generates sulfuric acid and is responsible for billions of dollars of damage each year to wastewater collection and treatment systems. It clearly and simply illustrates a problem that has long been recognized but has only recently become a serious widespread concern. It is therefore important to understand the basics of the "sewer sulfide cycle" before we can scientifically apply control measures.

Figure 2-7 is a basic schematic diagram of the sewer sulfide cycle that shows the important interactions between sulfur and the natural biochemistry in a sewer. Since the reactions of sulfur take place both in the water and in the air, the processes are divided into liquid phase and vapor phase. In this way, the diagram also takes on the general form of a gravity sewer with the liquid phase processes taking place at the bottom and the vapor phase processes on the top.

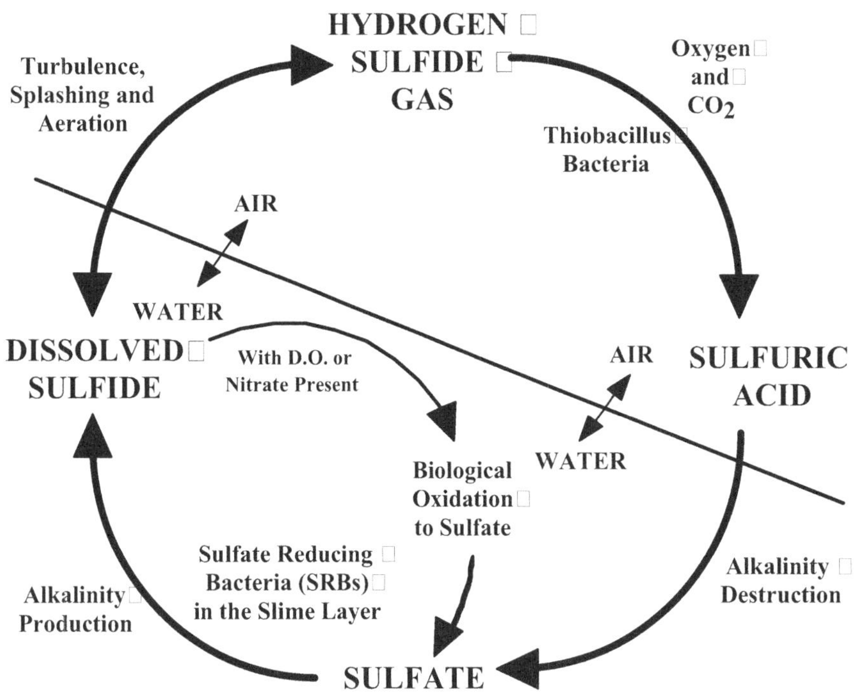

Figure 2-7. The Sewer Sulfide Cycle.

This sewer sulfide cycle effectively represents the chemical, physical, and biological processes occurring in sewers that involve all the various forms of sulfide. The actual biochemical processes it represents are complex and dependent upon many variables, but the process pathways illustrated in Figure 2-7 are common to all sewers.

There are four major segments of the cycle: the generation of sulfide, the release of sulfide, the conversion of hydrogen sulfide gas to sulfuric acid, and reaction of acid on sewer walls. Each segment comprises biological and chemical processes that fulfill that segment's particular function in the overall sulfide cycle. The four segments are linked by arrows representing the major pathways through which all sulfide processes must pass. These pathways provide us with an opportunity to interfere with the processes and thus reduce or stop the cycle of sulfide generation in sewers. The following describes each of the pathways in detail and illustrates how they can be used to control sulfide.

2.6.2.1 From Sulfate to Dissolved Sulfide

The cycle starts when sulfate ($SO_4^{=}$) in the wastewater is consumed by anoxic slime layer organisms to produce sulfide ($S^{=}$). Sulfide produced by the slime layer immediately attracts hydrogen atoms from water to form dissolved sulfide (HS^{-} + $H_2S(aq)$). When a hydrogen atom is taken from a water molecule it forms one hydroxide ion (OH^{-}). Hydroxide ions are alkaline since they can react with acids. Therefore, the creation of one hydrogen sulfide gas molecule generates alkalinity in the form of two hydroxide ions. This process occurs only underwater in the slime layer and the wastewater.

If the wastewater contains sufficient dissolved oxygen or nitrate, some or all of the sulfide can be reoxidized to sulfate to restart the process. This process occurs in the outer slime layer, which recycles the sulfur between sulfide and sulfate, as represented by the small arrow cycle back to "sulfate."

2.6.2.2 From Dissolved Sulfide to Hydrogen Sulfide Gas

The next pathway of the cycle represents the release of a dissolved gas ($H_2S(aq)$) into the sewer to exist as a free gas ($H_2S(g)$). It should be noted that as with almost all gasses that exist in nature, hydrogen sulfide gas would naturally be released from water containing the dissolved gas in an attempt to establish equilibrium with Henry's law. The rate of this release is dependent upon the surface area of the air-water interface.

For example, if we fill a beaker half-full with a solution containing exactly 1.0 mg/L of dissolved sulfide and then immediately seal it, some of the dissolved hydrogen sulfide gas would naturally be released from the water to exist as the free gas. Assuming that the pH of the solution remains neutral (7.0) and the dissolved sulfide concentration stays constant at 1.0 mg/L, the airspace in the beaker will eventually contain close to 300 ppm of hydrogen sulfide gas (well into the deadly range for hydrogen sulfide!). This concentration just happens to be the equilibrium point for hydrogen sulfide with Henry's law at 1.0 mg/L dissolved sulfide. If we place the sealed beaker in a quiet area and leave it totally undisturbed it may take many months or even years for this equilibrium to stabilize. However, if we vigorously shake the sealed beaker (again assuming that the pH of the solution remains neutral and the dissolved sulfide concentration stays constant at 1.0 mg/L), we will achieve our 300-ppm airspace concentration in about 30 seconds.

The reason we can cut the equilibrium time from several years to 30 seconds is surface area. In the undisturbed sample, the surface area of the air-water interface is fixed and easily determined. In the violently shaken sample, the air and water are mixed intensely so that all of the air exists as very small bubbles that are in constant contact with water. The combined surface area of the bubbles (total air-water interface) is many hundreds of times larger, which means that the equilibrium forces can now operate across the entire bubble-water surface area and reach equilibrium very quickly.

Note that this is the only pathway which has an arrow going in both directions. Although it rarely happens in sewers, if hydrogen sulfide gas concentrations are very high it could force some of the gas back into solution. The dispersing and dilution of gas throughout the sewer and the continuous production of sulfide by the slime layer will keep the dissolved concentration high enough to prevent reabsorption of the gas. The double-arrowed pathway represents this possible but unlikely event.

2.6.2.3 From Hydrogen Sulfide Gas to Sulfuric Acid

The next process of the sulfide cycle shows that hydrogen sulfide gas can be oxidized by *Thiobacillus* to produce sulfuric acid or can be released to the atmosphere. Once secreted by the bacteria on the sewer walls, the acid either reacts with the concrete and metal in the sewer to cause corrosion or runs down the sewer walls back into the wastewater where it is neutralized. Wherever atmospheric oxygen, carbon dioxide gas, water vapor and hydrogen sulfide gas (>2 ppm) are present, *Thiobacillus* bacteria can exist. This pathway occurs only on the abovewater surfaces in sewers.

2.6.2.4 From Sulfuric Acid to Sulfate

The last pathway completes one full cycle of transformation when unreacted sulfuric acid runs down the pipe wall and is neutralized in the wastewater flow. When it contacts the wastewater, the acid destroys the alkalinity (OH^-) that was produced when the original sulfate was reduced to sulfide. Therefore, there is no net change in wastewater pH from the generation of acid from hydrogen sulfide gas in sewers. It is also important to note that when sulfuric acid is neutralized by reacting with alkalinity in the sewage, sulfate is the product. This sulfate is once again available for uptake and conversion to hydrogen sulfide gas as shown. Through this ever-repeating, self-sustaining cycle, sulfur and sulfate are in a constant state of change in a sewer.

Figure 2-7 explains in basic terms the transformation pathways of sulfate and sulfide in sewers. However, it is most useful when technically evaluating and selecting sulfide control measures for a particular sewer system.

2.7 Using the Sulfide Cycle to Evaluate and Select Control Measures

The sulfide cycle and its pathways explain how one part of the cycle is dependent upon another. As mentioned above, we need to look at only the four major process pathways, each feeding the next step, to realize that interrupting one or more of these pathways could stop the cycle and thus stop sulfide odor and corrosion. By knowing how these pathways are related and some general information on sulfide control mechanisms, we can evaluate the various methods of sulfide control and prepare a short list of feasible alternatives.

All sulfide-related odor and corrosion control alternatives for collection systems fall into one of three basic control strategies: design and operational activities, chemical-liquid phase operations, and ventilation/vapor phase operations. Knowledge of the interdependent relationships of sulfate concentration, slime layer biochemistry, turbulence conditions, and *Thiobacillus* requirements allows an investigator to evaluate how any particular sulfide control alternative would work in any particular collection system.

As an example, let's assume that we want to interrupt the cycle in a 925-mm (36-inch) diameter gravity sewer where sulfate is biologically reduced to sulfide in the slime layer.

(Note that stopping sulfate reduction in the slime layer is an interesting place to start since it requires clarification of a small but important detail of sulfide generation in sewers: Technically, there is no way to completely stop the reduction of sulfate by the slime layer short of physical or chemical removal the slime layer itself. Even when wastewater contains high concentrations of dissolved oxygen, sulfate still penetrates through the aerobic zone to be reduced to sulfide. When this sulfide migrates back up through the aerobic zone, it is oxidized to sulfate with no net effect on wastewater sulfide concentrations. Although no sulfide is being released back to the wastewater from the aerobic zone on top of the slime layer, sulfide is still being generated in the zone beneath.)

2.7.1 Interrupting the First Pathway

In our example, we will evaluate ways to stop the first pathway (reduction of sulfate to sulfide), as illustrated in Figure 2-8. Table 2-2 lists by category the options available for interrupting this pathway.

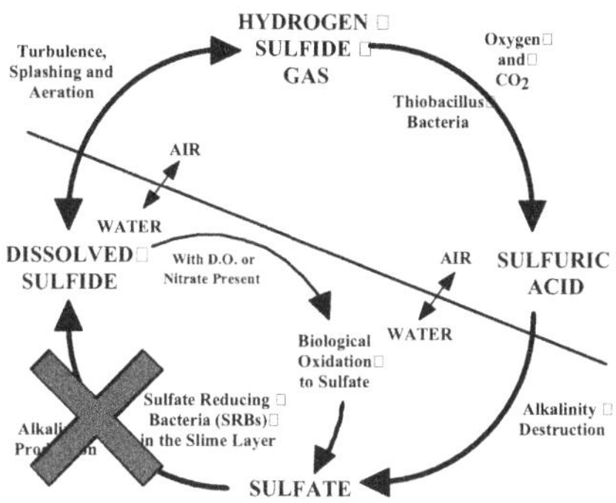

Figure 2-8. Interrupting the First Pathway of the Sulfide Cycle.

Table 2-2. Strategies for Interrupting the First Pathway.

Category	Strategy	Method(s)
Design and operational	Physically remove the slime layer	Pigging Water jetting
	Physically remove sulfate from wastewater	Ultrafiltration/nanofiltration (not practical) Sewer-use ordinances (regulate sulfate discharge)
Chemical and liquid phase	Chemically remove the slime layer	Caustic slug dosing
	Chemically stop sulfate reduction in the slime layer	Addition of anthroquinone (metabolic inhibitor)
	Chemically remove sulfate from wastewater	Barium precipitation (not practical)
Ventilation/vapor phase	None	None

2.7.1.1 Design and Operational Methods

In our example, physically removing the slime layer would stop sulfate reduction and can be accomplished by pigging in force mains or high-pressure water jetting in gravity sewers. Pigging involves forcing a "pig" the same diameter as the pipeline through the pipe to scrape off the soft biological slime layer. Pigging is not feasible in a gravity sewer, and high-pressure jetting of a sewer every few days would be cost prohibitive and might accelerate pipe deterioration. Neither of these physical removal methods is in widespread use due to high labor requirements and regrowth of the slime layer in 3-7 days.

Removing sulfate from wastewater means controlling sulfate in all discharges to a collection system, which is very difficult, if not impossible, because sulfate is found almost universally in all waters and is very soluble in most of its forms. One conceptual physical method of sulfate removal is to regulate or control the discharge of sulfate through sewer-use ordinances. There are only a few instances where sulfate discharges have been regulated, and in these cases the intent was to reduce extremely high discharge concentrations to levels closer to background. Sulfate can also be physically rejected using certain membrane processes (ultrafiltration and nanofiltration) but the process is expensive and not practicable on a large-scale basis.

2.7.1.2 Chemical and Liquid Phase Methods

Chemical removal of the slime layer can be accomplished through caustic shock dosing, a chemical method. By killing all of the bacteria that make up the slime layer, the reduction of sulfate is stopped. Neither physical nor chemical removal of the slime layer is used very much since it can grow back in 3-10 days, depending upon BOD and temperature.

Chemically stopping sulfate reduction can be accomplished by adding the chemical anthroquinone to wastewater. This chemical interferes with the metabolism of sulfate-reducing facultative organisms, rendering them useless. The chemical is not very soluble at typical wastewater pH, which limits its usefulness.

Barium is the only element which reacts with sulfate under normal conditions to form an insoluble precipitate. Barium is also toxic to humans and is not practicable for use in wastewater.

2.7.1.3 Summary

We have just evaluated all of the potential alternatives for preventing the reduction of sulfate to sulfide in our example sewer. Additional detailed knowledge of our sewer would provide the data necessary to put costs to feasible methods and arrive at the least costly alternative for stopping sulfate reduction in the slime layer. In actual practice, prevention of sulfate reduction is rarely considered or evaluated except in the case of a few very high sulfate concentration industrial discharges, and then only to reduce the concentration and not eliminate sulfate. Sulfate is generally assumed to be present in all wastewaters and it is rarely feasible to remove it.

2.7.2 Interrupting the Second Pathway

Let's now consider interrupting the second sulfide cycle pathway in our sewer. In this case, we are allowing sulfide to be generated in the slime layer but will try to prevent its release to the sewer atmosphere.

In this example we will evaluate ways to interrupt the second pathway of the sulfide cycle (Figure 2-9); Table 2-3 lists by category the options available for interrupting this pathway.

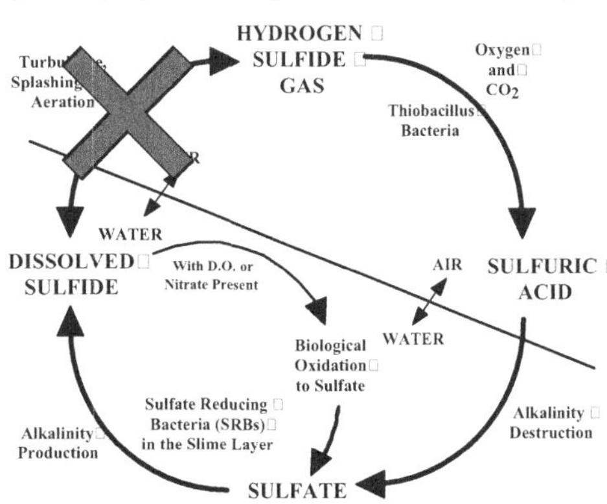

Figure 2-9. Interrupting the Second Pathway of the Sulfide Cycle.

Interrupting the second pathway of the sulfide cycle is much more common, and there are many potential alternatives which can do this. We must first list the alternatives to stop this pathway and categorize them once again into design and operational activities, chemical-liquid phase operations, and ventilation/vapor phase operations; most of the alternatives in this category are chemical and liquid phase methods and there are no ventilation alternatives to prevent the release. Ventilation controls become effective only after hydrogen sulfide gas is released in the next pathway of the cycle.

Table 2-3. Strategies for Interrupting the Second Pathway.

Category	Strategy	Method(s)
Design and operational	Physically prevent turbulence	Sewer velocities/hydraulic jumps Drop structure/drop manhole modifications Modify/design force main terminations for turbulence reduction Controlled release of sulfide from wastewater (stripping—requires vapor phase treatment for odor control)
Chemical and liquid phase	Oxygen injection/ addition	Air injection/diffusion Pure oxygen injection
	Chemical addition	Chemical oxidants (chlorine compounds, hydrogen peroxide, potassium permanganate) Precipitants (ferrous and ferric chloride and sulfate) pH adjustments (caustic addition, magnesium hydroxide addition) Bioactive oxidants (nitrate/nitrite)
Ventilation/vapor phase	None	None

As we can see in Table 2-3, it is in interrupting this pathway where most of the current sulfide controls operate. We saw above that preventing the transformation pathway of sulfate to sulfide in sewers is very difficult, if not impossible. Therefore, we must try to remove the sulfide immediately after it emerges from the slime layer but before it can be released to the air. Because of the high number of sulfide control methods focused on this pathway of the sulfide cycle, descriptions will be brief.

2.7.2.1 Design and Operational Methods

Since preventing sulfide release is the goal and since turbulence controls stripping, we could focus on turbulence reduction to reduce the release of sulfide. Sewers can be designed to avoid excessive slope/velocity changes and hydraulic jumps. Drop structures and manholes should be designed with drop pipes to handle wastewater smoothly and without agitation. Force mains should discharge below a standing body of water whenever possible to absorb the kinetic energy of the discharge without splashing or turbulence. At the outer range of possibilities is the intentional stripping of sulfide and removal from the air as hydrogen sulfide gas. Although stripping seems counter to the goal of preventing release of hydrogen sulfide gas, if the release can be accomplished in a location where facilities have been designed to handle the conditions, this effectively prevents release from downstream facilities.

2.7.2.2 Chemical and Liquid Phase Methods

The vast majority of chemical and liquid phase sulfide controls operate by interfering with this pathway. They range from the powerful oxidants such as chlorine, hydrogen peroxide, and potassium permanganate to the precipitants like iron that work to precipitate sulfide as a metal sulfide. Some chemicals take advantage of the pH dependence of sulfide by raising the pH of the entire wastewater flow. This shifts the sulfide equilibrium from the releasable dissolved gas form to the nonreleasable bisulfide ion, as illustrated in Figure 2-4. It is in this category that we also find oxygen and nitrate chemicals, which we discovered earlier form an aerobic or nitrate-dominated zone on top of the slime layer.

There are many chemicals and liquid phase treatments available to interfere with this cycle pathway in many different ways. It would not be appropriate to try to discuss each of them here. A full technical description and analysis of each of these chemical treatments is discussed in detail in Chapter 7.

2.7.2.3 Summary

Controlling the second sulfide cycle pathway is the exclusive domain of chemical treatments and other liquid phase processes. Chemicals are the most common sulfide control technique and are used primarily to control odor complaints and not corrosion.

2.7.3 Interrupting the Third Pathway

To continue our example, let's now consider that we want to interrupt the third cycle pathway in our sewer. In this case, we are allowing sulfate to be reduced in the slime layer and released to the sewer atmosphere but will try to prevent its conversion to sulfuric acid.

In this example we are essentially interrupting the third pathway of the sulfide cycle (Figure 2-10); Table 2-4 lists the options available for interrupting this pathway.

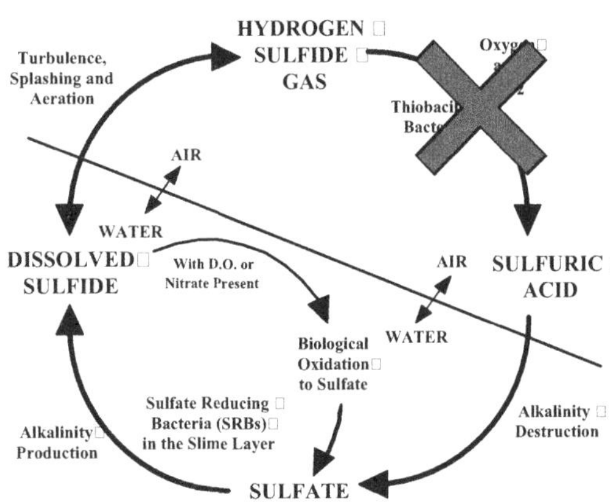

Figure 2-10. Interrupting the Third Pathway of the Sulfide Cycle.

Interrupting the third pathway of the sulfide cycle involves preventing hydrogen sulfide gas from being oxidized to sulfuric acid. This means that we need to stop or limit the growth of *Thiobacillus* bacteria or otherwise interfere with the uptake of hydrogen sulfide gas and generation of acid.

Once again, if the alternatives to stop the sulfide cycle at this pathway are listed (with the exception of preventing release of H_2S discussed above) it is evident that there are possibilities in two of the three categories: design and operational and ventilation/vapor phase alternatives but no chemical-liquid phase controls. This is because acid generation occurs on the abovewater surfaces in sewers and there is nothing that can be added to the wastewater that will affect this area of sewers.

Table 2-4. Strategies for Interrupting the Third Pathway.

Category	Strategy	Method
Design and operational	Physically prevent acid generation	Surcharge sewers
Chemical and liquid phase	Chemically prevent acid damage	Crown spraying
Ventilation/vapor phase	Modify sewer environment to prevent *Thiobacillus* growth	Ventilate sewer and treat odorous air

2.7.3.1 Design and Operational Methods

One way to prevent oxidation of H_2S to sulfuric acid is to prevent the growth of *Thiobacillus* bacteria. Since these bacteria require three gasses in combination with high humidity and/or moist growing surfaces, we can try to eliminate one of these factors. Eliminating gasses from the sewer would require operating the sewer with no available headspace for air or required gasses like H_2S or CO_2. By operationally surcharging the sewer, the bacteria to not have access to the necessary gasses, and thus corrosion is effectively stopped. Caution should be given to "crown cutting," which affects surcharged pipes as discussed below.

It would typically be assumed that *Thiobacillus* corrosion would not be present in a full-flowing gravity sewer since full pipes typically do not allow gasses to be present. However, a phenomenon called "crown cutting" or "crown corrosion" has been found in concrete pipes which intentionally flow full. The process starts when entrained air bubbles in the flow coalesce on the pipe crown in a place where the bubble can remain stationary against hydraulic forces. This is typically at a joint in reinforced concrete pipe (RCP) or any high spot in the line where hydraulic shear cannot dislodge the air bubble. All of the conditions required for *Thiobacillus* will exist inside this bubble of air. As acid is generated on the above-water portion of this air pocket, the concrete corrodes upward, which makes the air pocket larger. Over time, acid production continues on the top surface and can even breach the crown of the pipe. The corrosion does not extend laterally beyond the limits of the bubble since it is continuously wet and not exposed to gasses. Figure 2-11 is a photograph of crown corrosion in a full-flowing concrete pipe.

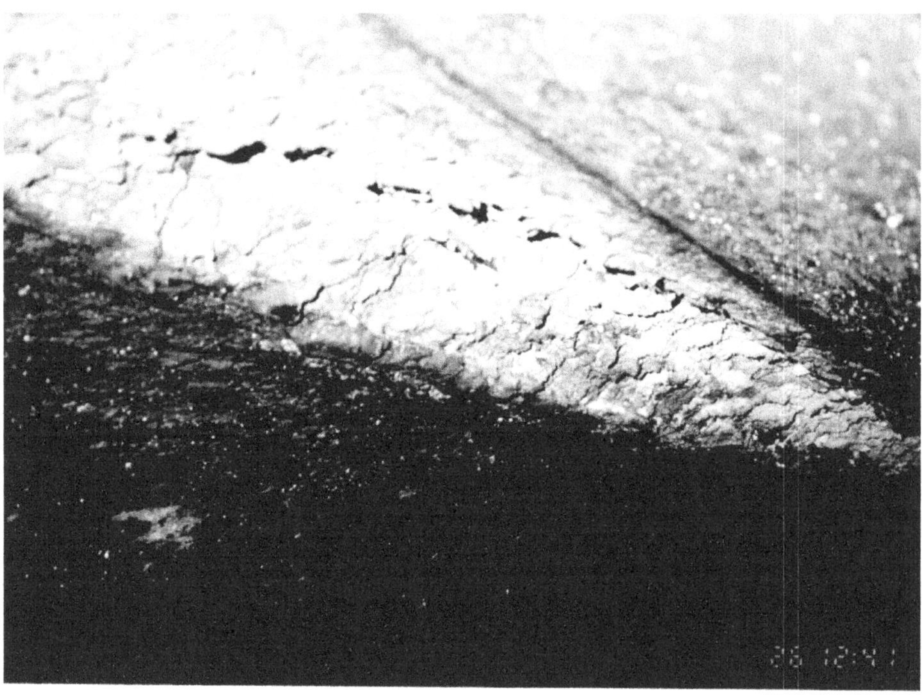

Figure 2-11. Crown Corrosion.

Crown spraying typically involves spraying a caustic or alkaline chemical on the exposed surfaces of sewers and structures to kill *Thiobacillus* bacteria and neutralize any acid that may be on the surface. (This process is discussed more in Section 2.7.4.) However, crown spraying (most often used with magnesium hydroxide slurry) also functions to prevent the uptake of hydrogen sulfide gas by *Thiobacillus* bacteria. Floating "skids" or "boats" are used to transport a spray nozzle assembly down the sewer while pulling chemical supply hoses along and spraying slurry as it goes. The process is usually applied from one upstream manhole to the next downstream manhole and requires two crews. The skid is pulled through the sewer so that speed and therefore application rate can be controlled. The thick layer of alkaline slurry covers the above-water surfaces of sewers, effectively cutting off the food supply for *Thiobacillus*. In addition, the slurry reacts directly with hydrogen sulfide gas in the sewer to form a magnesium sulfide complex (blue or black in color). The high pH of the slurry surface also prevents *Thiobacillus* growth until the pH is naturally reduced to acceptable limits.

2.7.3.2 Ventilation/Vapor Phase Methods

The forced ventilation of sewers can be used to reduce the rate of corrosion and to stop corrosion in certain situations. Extracting air from a sewer with a fan will cause a similar volume of air to enter the system. This dilutes the concentration of hydrogen sulfide gas in the sewer which in turn reduces the mass of sulfuric acid being generated and thus reduces the corrosion impact. Similarly, the influx of outside air and its subsequent withdrawal will reduce the relative humidity in the sewer, which will change the moisture levels in the sewer and dry some of the pipe surface. A dry surface cannot support *Thiobacillus* bacteria, so sulfuric acid will not be produced. Additionally, the exchange of air will also change the air temperature in the sewer, which will affect the dewpoint and condensation on the interior sewer walls. All of these actions will serve to reduce the rate and extent of corrosion. It is important to point out that ventilation alone will not be as effective as when used in conjunction with one or more other control methods. Ventilation will never be able to stop corrosion completely.

2.7.3.3 Summary

Very few of the common sulfide control methods act to interfere with the generation of sulfuric acid from hydrogen sulfide gas. Those that do can rarely be 100% effective.

2.7.4 Interrupting the Fourth Pathway

Let's now consider that we want to interrupt the fourth pathway of the cycle in our sewer. In this case, we are allowing sulfate to be reduced in the slime layer, released to the sewer atmosphere, and converted to sulfuric acid but will try to prevent corrosion of materials before it runs back into the wastewater.

In this example the researchers evaluated ways to interrupt the fourth pathway of the sulfide cycle (Figure 2-12); Table 2-5 lists by category the options available for interrupting this pathway.

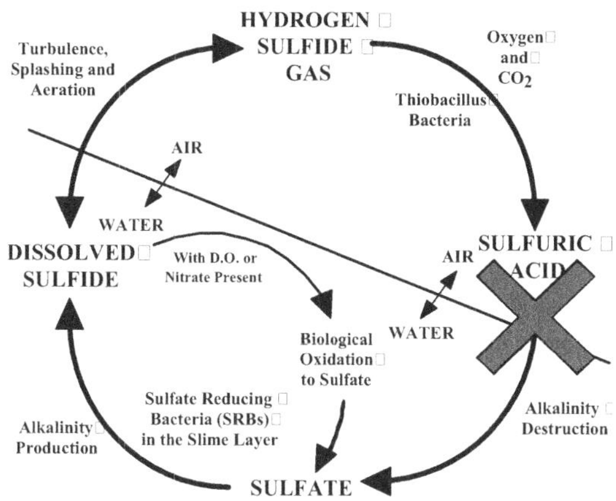

Figure 2-12. Interrupting the Fourth Pathway of the Sulfide Cycle.

Interrupting the fourth pathway of the sulfide cycle involves preventing sulfuric acid from contacting sensitive surfaces. Using this pathway for control of sulfide involves protecting sensitive corrosion-prone surfaces. This latter method could be used without odor treatment systems if the location is remote and there is no potential for complaints, although this situation is unlikely.

Once again, if the alternatives to stop this pathway are listed, it is evident that there are design and operational alternatives but no chemical and liquid phase or ventilation/vapor phase controls. This is because acid generation occurs on the abovewater surfaces in sewers and there is nothing that can be added to the wastewater that will affect this area of sewers. Ventilation controls prevent oxidation of hydrogen sulfide to sulfuric acid through interruption of the third pathway only.

Table 2-5. Strategies for Interrupting the Fourth Pathway.

Category	Strategy	Method
Design and operational	Use of inert construction materials	Plastics (PVC, HDPE, PE, PP, FRP) Coatings on concrete Cast-in-place liners on concrete (PVC, HDPE) Applied liners on concrete (PVC)
	Crown spray	Periodically spray crowns of sewers with caustic chemicals (sodium hydroxide, magnesium hydroxide)
Chemical and liquid phase	Crown spray	Periodically spray crowns of sewers with caustic chemicals (sodium hydroxide, magnesium hydroxide)
Ventilation/vapor phase	None	None

2.7.4.1 Design and Operational Methods

One way to prevent corrosion from even starting in the first place is to use acid-proof materials of construction. Selecting the proper pipe material for a new sewer must consider corrosion impacts. When corrosion resistance alone is considered, pipe material selection favors plastic materials such PE/HDPE, PVC, or various resin-based, composite materials such as polymer concrete, GRP, or FRP.

Inert pipe materials are still economical for pipe diameters less than about 72 inches, which comprises the vast majority of the total service area of the typical collection system. The vast number of plastic pipe types, formulations, and configurations makes a complete discussion in this chapter impossible. However, care should be exercised to determine the specific formulation of plastic pipes since all plastics are not equal. For instance, PVC pipe is manufactured using a number of different filler materials. This is due to the cost of the resins used and the strength of each particular resin. Fillers are added to provide bulk and mass to the pipe and reduce excessive resin consumption. Some filler materials can be attacked by acidic conditions, which leach the fillers from the pipe and reduce the structural capacity.

Coatings. By far the most common method used to interrupt this pathway is protecting sensitive sewer surfaces from contact with acid. The most common protection method (also the one with the shortest effective lifetime) is applying a coating, specifically, a product applied to a concrete surface as a liquid or semisolid and then allowed to chemically cure. Coatings used in the wastewater industry are generally either epoxies or urethanes. Many different formulations of each are available. Coating formulations should be selected based upon the requirements of the service conditions and performance characteristics of the coating product. Coating products must be evaluated for use on a case-by-case basis. The successful application of coatings is very environmentally sensitive. It can be extremely difficult to successfully apply coatings under the temperature, humidity, and flowing water conditions found in sewers. These factors often cause microscopic holes (pinholes) to form in the coating even when it is applied under strict supervision and inspection. Pinholes and other defects found in coatings provide a direct route for acid produced on the surface to reach the concrete and cause attachment failure.

Liners. When the cost of producing such large-diameter (over 72 inches) inert material pipes is considered, it is quickly found that the cheapest material by far is reinforced concrete pipe (RCP). The relatively low cost and high-strength of concrete makes it economically attractive for large-diameter, deep sewers, but the corrosion resistance of concrete is minimal. Concrete is readily attacked and destroyed by strong mineral acids like H_2SO_4; hence the need for liners.

Cast-in-place liners have knobs, Ts, ribs, or some other type of protrusion from one side of the liner sheet that are embedded into concrete or grout during construction. When the liner is attached to the concrete formwork with the protrusions facing outward, the opposite smooth, acid-resistant side of the liner is exposed when the forms are removed. Brand-name examples are T-Lock (PVC) and Bekaplast, Studliner, and Agru Sure-Grip (all PE). These types of liners can be used for either new construction or rehabilitation, and although new construction is where they are most often used, some of these liners are gaining acceptance as viable rehabilitation methods due to advances in grouting technology and in-pipe construction techniques. Whether cast-in-place liners are used for new or rehabilitation service, they relieve hydrostatic and vapor pressure by providing a pathway for gas and water escape into the sewer without sacrificing corrosion protection. Since the anchors are the only part of the liner embedded in the pipe wall, the flat surface of the liner is held close against the concrete but not tightly adhered to it. This allows "weep water" and vapor pressure to equalize into the sewer between the embedments without creating an open airspace where gasses can accumulate. Figure 2-13 shows a typical T-type cast-in-place liner.

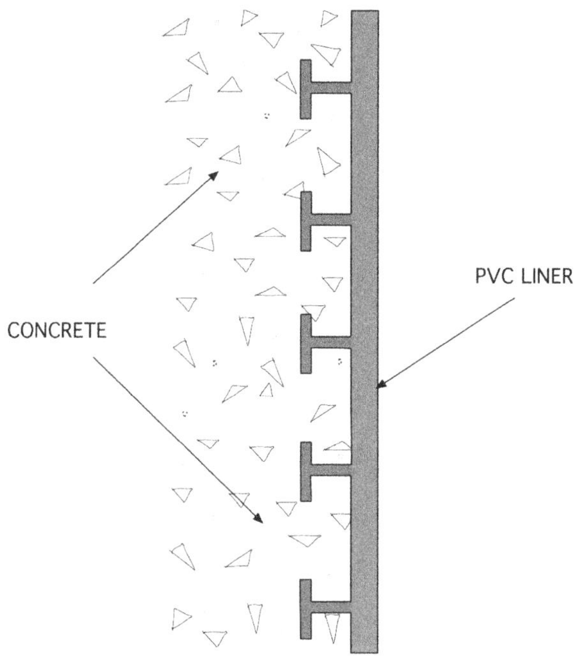

Figure 2-13. Typical Cast-in-Place Liner Profile on Wall.

Mechanically attached liners can be almost any material, although rigid (unplasticized) PVC and HDPE are most common. The most significant variation is the method of attachment, which can vary from small expansion-type anchors to stainless steel batten strips with epoxy grout anchors, as shown in Figure 2-14. The high degree of flexibility offered by mechanically attached liners makes them adaptable to a wide variety of conditions. Attachment systems can be designed to safely accommodate the anticipated hydraulic and hydrostatic forces. Liner thicknesses can also be easily varied for additional hoop strength to adjust to specific design needs. These liners have been used primarily for new construction, although numerous rehabilitations have been reported with equal success. Mechanically attached liners, like the cast-in-place liners, do not resist hydrostatic and vapor forces but pass them harmlessly to the sewer environment. Like the cast-in-place liners, the flat inner surface of the liner is held closely against the interior of the concrete pipe but not fastened to it. This allows water to flow harmlessly into the sewage and not create a localized high-pressure area, which could create a bulge and make the liner fail. Due to high flow failures in large-diameter sewers, mechanically attached liners are now used almost exclusively in structure protection where hydraulic forces can be controlled.

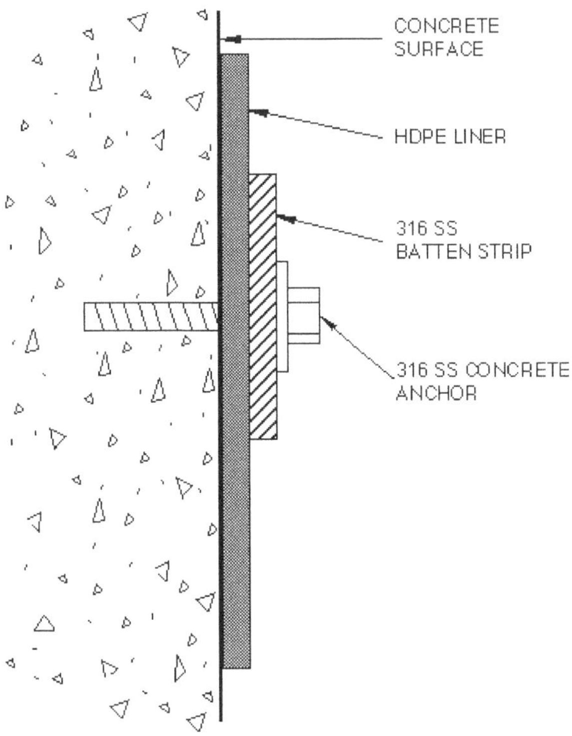

Figure 2-14. Mechanically Attached Liner Detail.

Chemically attached liners come in two basic forms: One type has protrusions on one side of a PVC sheet that are pressed into freshly applied epoxy mastic on the concrete. The mastic flows around the protrusions to provide secure attachment when cured. Another form of chemically attached liner is a flat PVC sheet that is pressed against a freshly applied polyurethane material as illustrated in Figure 2-15. A chemical additive on the PVC sheet causes chemical fusion and cross-polymerization between the urethane and the PVC, resulting in a very secure bond. These liners can be applied over new concrete; however, they are more commonly used in concrete rehabilitation projects. There are other, more-secure and more-economical methods for new construction than chemically attached liners. One of the potential pitfalls of a chemically attached liner is the requirement that it physically resist *all* hydrostatic and vapor pressure forces trying to push it off the wall. Since the liner is "glued" to the pipe wall with an epoxy or urethane mastic, the attachment of the epoxy or urethane to the pipe wall is critical to liner performance. The ability to resist anticipated hydrostatic and vapor pressures should be adequately demonstrated.

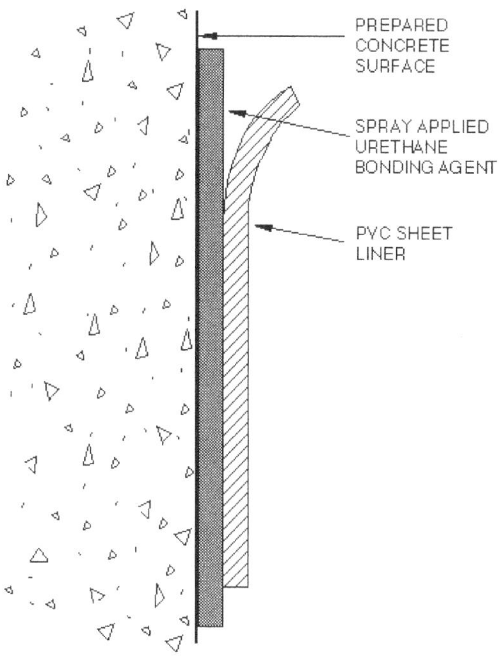

Figure 2-15. Typical Chemically Attached Liner.

Crown Spraying. This process involves spraying a caustic or alkaline chemical on the exposed surfaces of sewers and structures to kill *Thiobacillus* bacteria and neutralize any acid that may be on the surface. Floating "skids" or "boats" are used to transport a spray nozzle assembly down the sewer while pulling chemical supply hoses along. The process is usually applied from one upstream manhole to the next downstream manhole and requires two crews. The skid is pulled through the sewer so that speed and therefore application rate can be controlled. Sodium hydroxide was first used for this purpose, but it is hazardous to handle. This chemical has the consistency of water and any excess runs down into the wastewater. Magnesium hydroxide has been used more recently to provide the same effect without hazardous concerns. Magnesium hydroxide is also applied as a thick slurry, which puts more alkalinity on the crown and keeps it there. Depending upon the rate of corrosion and the condition of the pipe, crown spraying can cost effectively delay sewer rehab or prolong the life of pipes almost indefinitely.

2.7.4.2 Summary

The prevention of acid contact is the exclusive domain of inert materials, liners and coatings. When considering corrosion protection for new sewers, the increase in sulfide-related corrosion should be anticipated to continue and even more corrosive conditions will be experienced in the future. This would tend to favor inert materials and cast-in-place liners.

2.8 Corrosion Control Guidelines for Gravity Sewers, Force Mains, and Wastewater Collection Systems

Because wastewater flow itself plays a role in odor and corrosion within collection systems, those systems designed to reduce turbulence also minimize the effects of hydrogen sulfide, corrosion, and sewer ventilation in wastewater collection systems. The following design guidelines should be used prior to and during preparation of the engineering alternatives analysis and design for wastewater facilities.

2.8.1 Sulfide Control Design Considerations

The following guidelines help minimize the production of dissolved sulfide and the release of hydrogen sulfide gas into wastewater collection systems.

2.8.1.1 Gravity Sewers

Sufficient wastewater velocity, if maintained in gravity sewers, can minimize the production of dissolved sulfide. An absolute minimum velocity of 2.0 fps at low flow is recommended to reduce the accumulation of sewer solids in the pipe and allow potential reaeration of the wastewater to prevent excessive production of sulfide.

Excessive velocities—generally those over 6.0 fps—cause turbulence in the wastewater and strip hydrogen sulfide gas and should be avoided, particularly downstream of sulfide sources such as force main discharges. The sulfide concentration in the wastewater to be conveyed by the gravity sewer should be considered before designing for velocities that could cause the stripping of hydrogen sulfide gas.

Turbulence at points of intercepted flow (manholes or wyes) should be minimized by smooth transitions. Hydraulic drops and outfalls should be avoided. The use of drop manholes should be minimized, and optional methods such as submerged-outlet drop pipes or contained drops should be considered for outfalls up to 20 feet to contain and minimize turbulence (Figures 2-16 and 2-17).

Avoid combining wastewater flows at a confluence angle greater than 90 degrees, and never join sewers with opposing flows (confluence angle of 180 degrees). Confluence angles of 45 degrees or less have been found to reduce sewer pressurization, turbulence, and corrosion (Figure 2-18).

Sudden or dramatic changes in direction (such as 90-degree bends) and sudden changes in slope or velocity should be avoided due to turbulence, pressurization, and ventilation concerns. The longest radius horizontal curves possible should be used.

Inverted siphons should be avoided whenever possible. When absolutely necessary, inverted siphons should be provided with adequate air jumper or odor treatment capabilities and positive corrosion control measures such as liners or inert materials.

Never allow a gravity sewer's diameter to decrease in the downstream direction even if hydraulically correct. This causes turbulence, dramatic ventilation effects, localized pressurization, and odor release. More money will be spent to correct the odor and corrosion problem than was saved in pipe material and construction costs.

2.8.1.2 Pump Stations and Force Mains

The use of pump stations and force mains should be avoided whenever gravity sewer designs are feasible and constructible.

Pump stations and force mains should be designed to minimize hydraulic detention time within the pipeline, thereby reducing dissolved sulfide production.

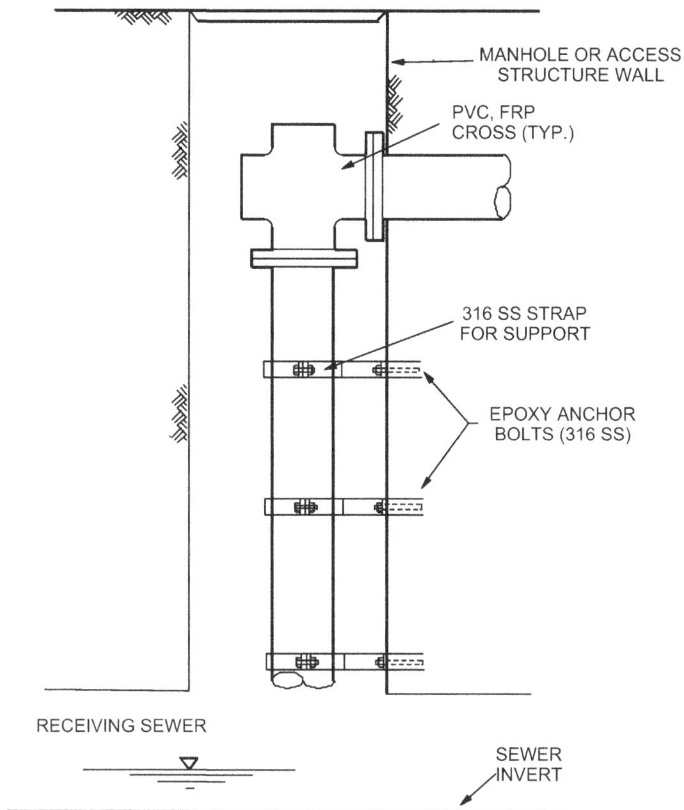

Figure 2-16. Inside Drop Structure to Minimize Turbulence of Gravity Drops.

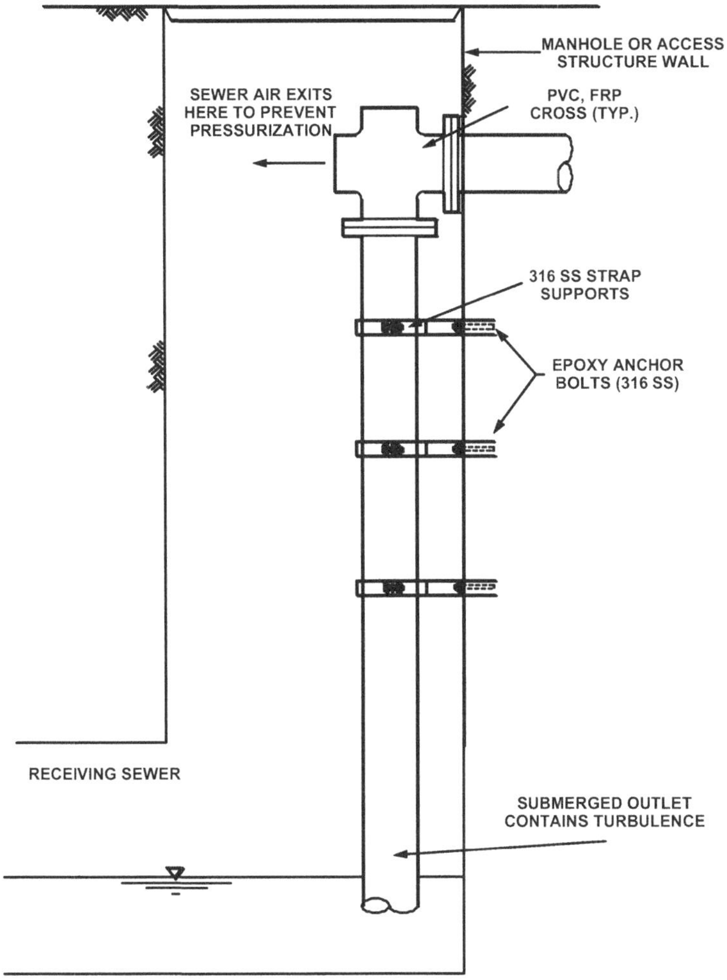

Figure 2-17. Submerged Inside Drop Structure.

Force mains should be designed so that discharges directly into a wet well minimize turbulence and the stripping of hydrogen sulfide gas. Wherever possible, discharges should be submerged to completely absorb hydraulic energy (Figures 2-18 through 2-24). Back-siphoning tendency should be evaluated before implementing.

Gravity sewer entrances into wet wells should be designed to minimize free fall and turbulence. Turbulence reduction measures similar to those in Figures 2-22 and 2-23 can be used if adjusting water levels and sewer elevations are impractical. Pump stations should be designed so that free air movement through the sewer is not impeded by wet well water levels except at flows greater than peak dry weather flow.

The number of air/vacuum release points for force mains should be minimized through grade and slope control. Avoid partially full force main conditions at any point in the pipe. Force mains should always have a flat or continuously rising profile, never a downward profile or negative slope.

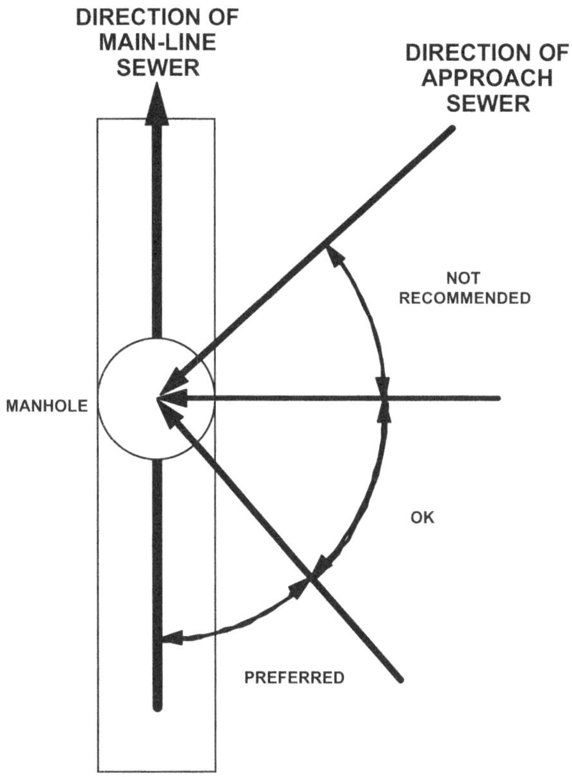

Figure 2-18. Sewer Approach Angles for Reducing Turbulence and Controlling Ventilation.

Consider the use of dual force main piping for new pump stations to maintain minimum velocities, provide for future capacity, allow redundancy of facilities, and reduce dissolved sulfide generation during early use years (if force mains can be drained by gravity).

Consider also programming the pump station controller to periodically surcharge the wet well to high levels and then call for all pumps to scour the slime layer from the pipe. This will minimize sulfide generation in the force main if done every three days.

Where the force main profile cannot be controlled and air/vacuum releases must be used, consider using variable-frequency drives on pump motors to maintain flow conditions and avoid excessive valve operation.

2.8.2 Corrosion Protection Design Considerations

The following guidelines have been prepared to assist in the reduction of corrosion potential in wastewater collection and treatment systems caused by exposure to hydrogen sulfide gas. In general, hydrogen sulfide gas concentrations below 2.0 ppm are not considered corrosive. Hydrogen sulfide gas concentrations between 2.0 and 5.0 are considered potentially corrosive, and sewers with concentrations averaging over 5.0 ppm should be 100% protected.

2.8.2.1 Gravity Sewers

Pipe material and pipe liners used for wastewater collection system construction should be unaffected by contact with sulfuric acid solutions of pH 0.5 (approximately 7% sulfuric acid by weight).

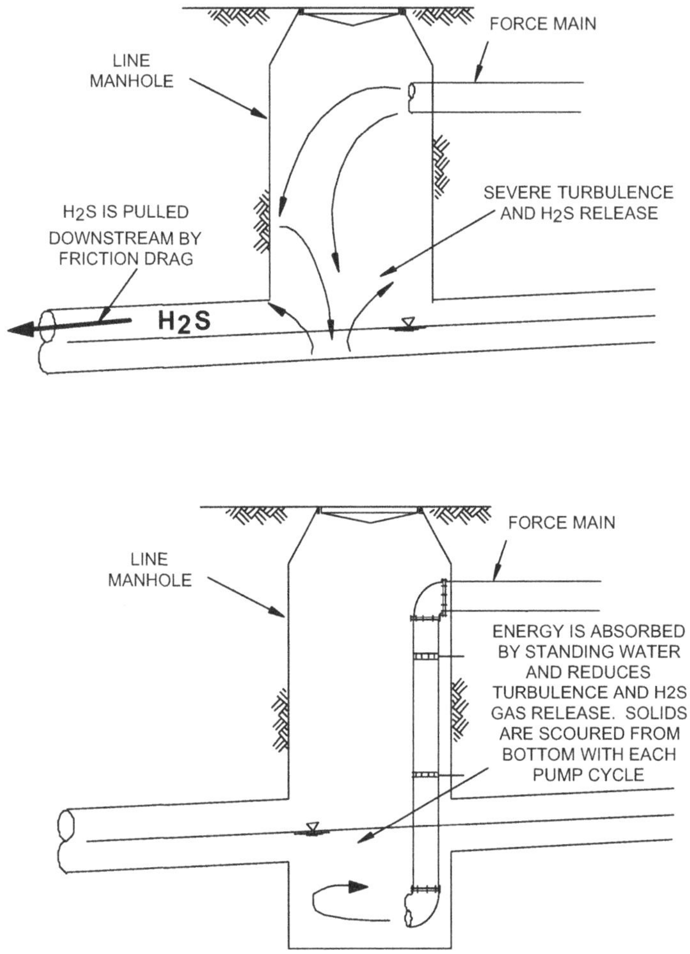

Figure 2-19. Force Main Turbulence: Before and After Reduction Using Modified Manholes.

Cementicious concrete products of any type (calcium silicate, calcium aluminate) should not be used for the conveyance of wastewater without an appropriately designed corrosion protection system (liner). Coatings cannot provide the service life of liners, and coated pipes will require recoating periodically. Coatings are not recommended for concrete that cannot be readily visually inspected and easily repaired without entering the pipe.

Particular attention should be given to corrosion protection of existing unprotected concrete gravity sewer pipes and manholes in areas of increased turbulence (intercepted flows, 90-degree bends, slope changes, drop manholes, and force main discharge locations) for a distance of 300 sewer diameters downstream and 50 sewer diameters upstream.

2.8.2.2 Pump Stations and Force Mains

Pump station wet wells should be constructed of or lined with inert materials. Plastic liners provide a longer life than applied coatings. Exposed, unprotected concrete of any kind should not be used in pump station wet wells or other areas exposed to average concentrations of hydrogen sulfide gas greater than 5.0 ppm.

Piping, valves, and fittings used inside pump station wet wells or areas exposed to significant concentrations of hydrogen sulfide should be constructed of, or lined with, inert materials. The only metal allowed is 316 stainless steel. Coated ductile iron is not acceptable.

Force mains should be constructed of inert pipe materials or factory-applied polyethylene-lined ductile iron pipe. Prestressed concrete cylinder pipe is not recommended for wastewater force main applications.

Pay particular attention to protecting structures downstream of turbulent force main discharge areas. Force main discharge manholes should be constructed from T-Lock concrete or inert materials only.

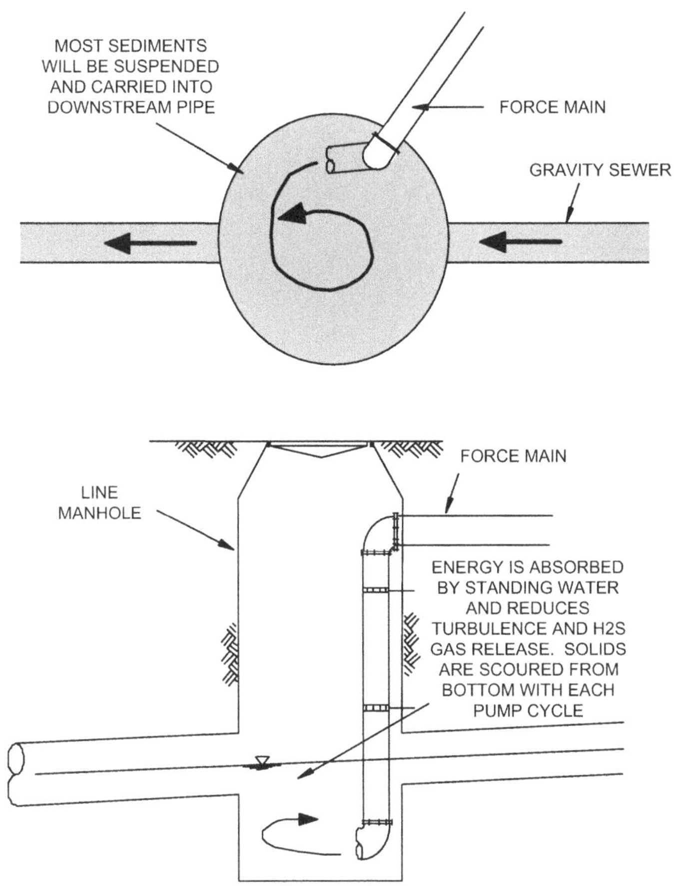

Figure 2-20. Force Main Line Manhole Turbulence Correction (Plan and Section Views).

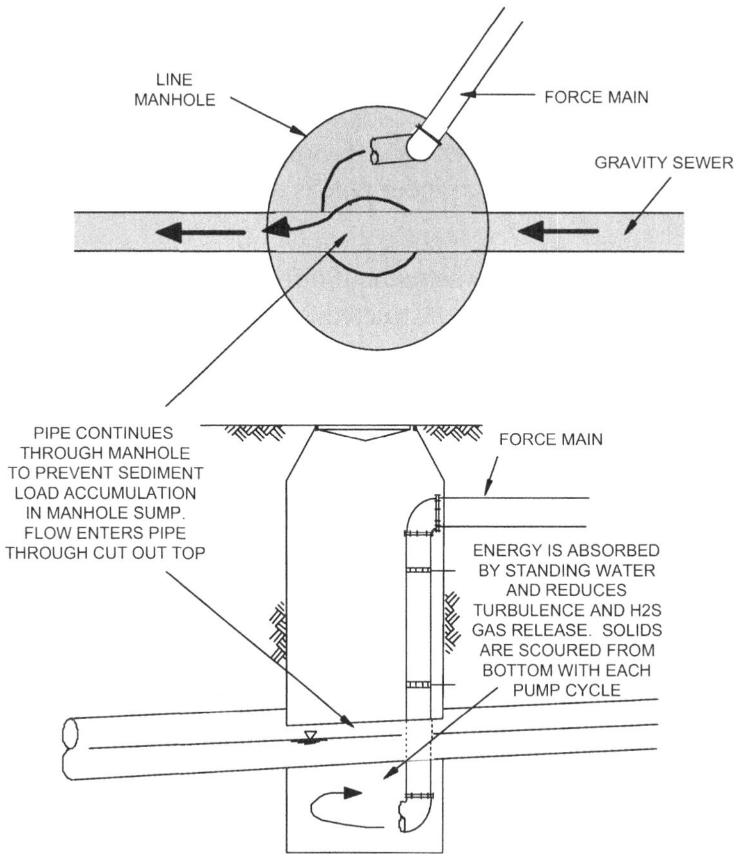

Figure 2-21. Line Manhole Turbulence Reduction Using Modified Manhole with Receiving Sewer Elevated (Plan and Section Views).

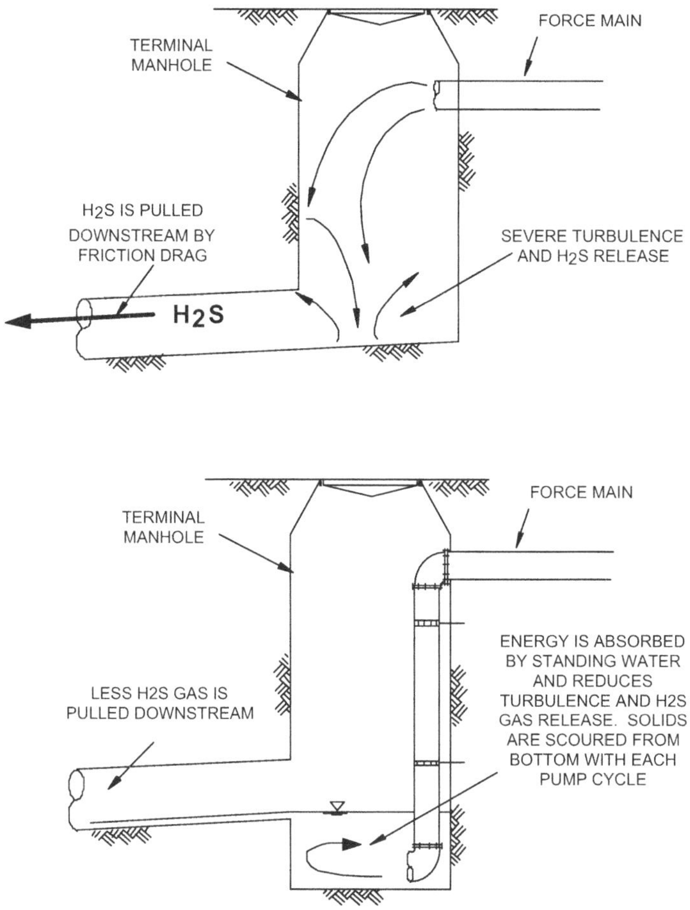

Figure 2-22. Terminal Manhole Force Main Connection Turbulence Control.

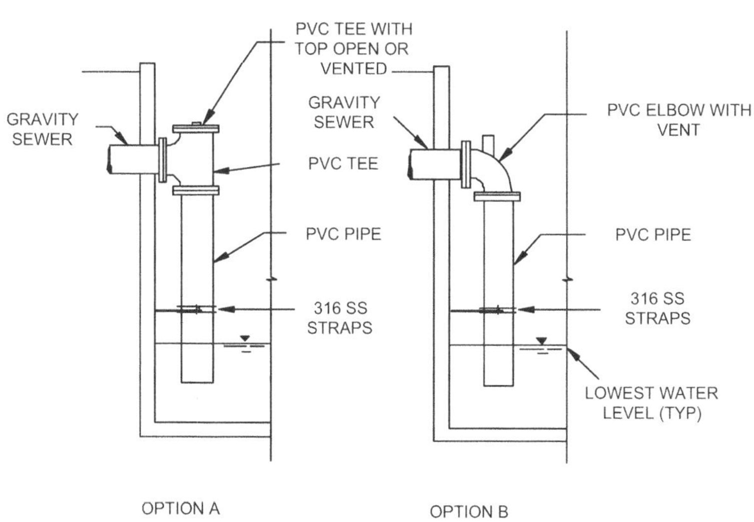

Figure 2-23. Conceptual Wet Well Turbulence Reduction by Modification of Influent Gravity Sewers.

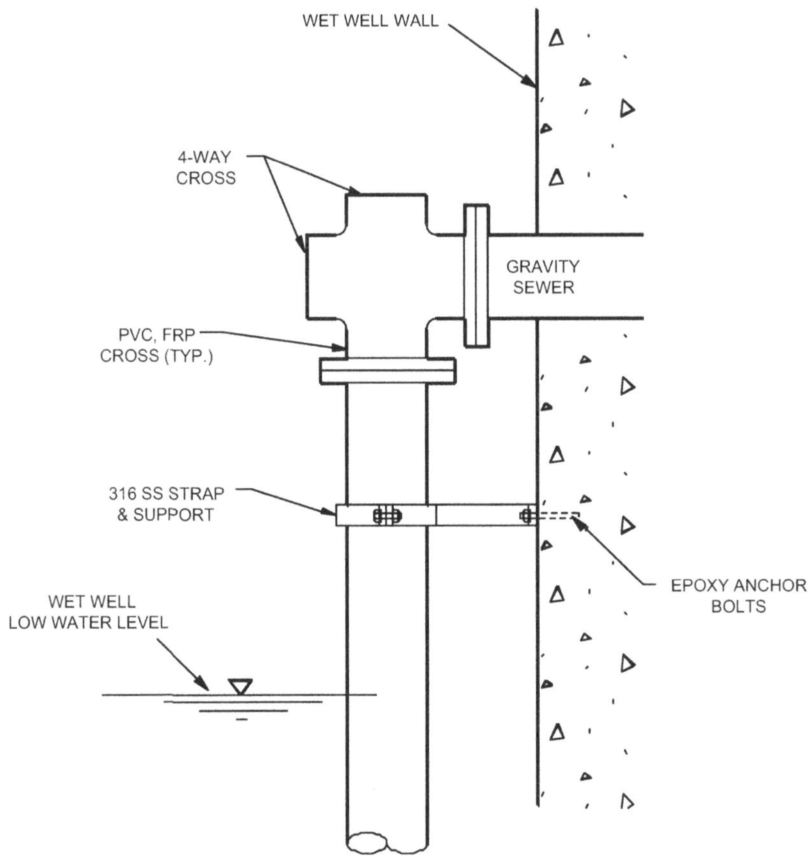

Figure 2-24. Wet Well Influent Sewer Turbulence Control Using Drop Pipe with Vent and Maintenance Access (Cross).

2.9 Summary

This chapter has explained what causes odors and corrosion in sewers, some of the important characteristics of the specific chemicals involved, and some of the methods available for interrupting the sulfide cycle. It has explained the general trends in wastewater sulfide generation and why the subject of collection system odor and corrosion control are important. The researchers also explained the basics of using the sulfide cycle to evaluate control mechanisms. As the researchers have seen, some of the most common methods for sulfide odor and corrosion control would be rejected from use when evaluated with the sulfide cycle. The information contained in the remainder of this manual expands upon these control methods to provide the information necessary to evaluate any collection system and select viable sulfide control mechanisms.

CHAPTER 3.0

LITERATURE SEARCH AND REVIEW

3.1 Introduction

The researchers used a systematic process to identify, obtain, and review current literature on wastewater collection system odor and corrosion. First, protocols for searching the published technical literature, unpublished documents ("gray literature") containing public and private organizations' information and experience, and literature included in previous WERF studies were developed. Second, the references obtained following the search protocol were compiled and screened for inclusion. After these papers were acquired, the research team cataloged and distributed them by topic category to project team experts for review. Finally, team members' literature reviews were completed and essential information about each paper recorded electronically. The literature search and review process was comprehensive and thorough and resulted in an annotated library of current literature relevant to the project topic.

3.2 Published Literature Search Protocol

The plan for conducting the published literature search consisted of identifying keywords and databases and dividing members into search teams. Project team members generated a comprehensive list of keywords (see Appendix A) covering each of 14 major topic categories within the broader topic of wastewater collection odor and corrosion. Those keywords were sorted into four major topic groups in order to share the work of carrying out the search. The topic groups and search teams were organized as follows:

- Group 1, Jeff Weis and Los Angles County Sanitation District (LACSD), searched collection system components and corrosion.
- Group 2, Shabbir Basrai and Orange County Sanitation District (OCSD), searched containment, ventilation, and gas phase treatment.
- Group 3, Dr. Richard Stuetz of the University of New South Wales and Joseph Cesca of CH2M HILL, searched liquid phase and biological treatment.
- Group 4, Dr. Richard Corsi of the University of Texas at Austin and Matthew Ward of CH2M HILL, searched emissions modeling and compounds.

The rationale for dividing the search was that different teams would have different resources; the teams could focus their efforts, minimize duplication of effort, and thereby explore more deeply the available literature abstracts. Some redundancy was built in to the process: published papers missed by one group could be caught by another. Each group searched a different set of keywords, but flagged all papers related to any relevant project topic. Some overlap occurred without causing unwarranted duplication of effort.

All English-language papers published during or later than 2000 were included. However, the search built on a previous WERF study that included a systematic search of papers published from 1990 to 2000 (with discretionary inclusion of earlier papers). Other papers published before 2000 were included at the discretion of each search team based on relevance to the research topic. A core list of bibliographic databases and search tools was selected according to the experience of the academic project team members. The selected databases were chosen to provide comprehensive coverage of English-language technical literature related to the project topic. A University of Texas reference librarian was consulted to verify that the most appropriate databases were chosen. In addition to the core databases, supplementary databases were used at the discretion of each search team. Table 3-1 lists the databases used.

Table 3-1. Databases and Indexes Queried for the Published-Literature Search.

Group	Published Literature Search Tool
Core (all groups)	Academic Search Premier
	Chemical Abstracts
	Engineering Index (Compendex)
	Dissertation Abstracts
Supplementary Group 1	Environmental Engineering Abstracts
	Civil Engineering Abstracts
	Pollution Abstracts
	Corrosion Abstracts
	Wilson Applied Science and Technology Abstracts
Supplementary Group 2	Applied Science full text
Supplementary Group 3	GEOBASE
	MEDLINE
	CAplus
	MEDLINE
	CAS REGISTRY
	CASREACT
	Proquest Science Journals
	Civil Engineering Database
	Proquest Digital Dissertation Database
	Science Citation Index Expanded
	Applied Science and Technology Abstracts
Supplementary Group 4	Science Citation Index Expanded

A three-pronged quality assurance/quality control (QA/QC) plan was adopted to prevent important published literature from being overlooked. First, the search teams queried alternative databases and indexes in order to find papers not identified by the core databases. Second, during the search process a database of included papers was maintained and made available to all project team members. Each project team member identified, on the basis of professional experience, important literature missed by the keyword search. Third, Googlescholar, a broad, multidisciplinary academic search engine, was queried to pick up any papers that had been missed through the other search methods. For the Googlescholar search, a less restrictive keyword search string was employed in order to identify all papers related to any of the topic categories.

3.3 Gray Literature Search Protocol

So that the project could benefit from public utilities' and private vendors' knowledge and experience, unpublished, or gray, literature was solicited from an extensive list of public and private organizations. A total of 149 public utilities were sent a formal request for gray literature.

The letter sent to each utility requested a list of studies and technical papers, owned by the utility, of potential interest to the project. Upon receiving lists from responding utilities, the team requested that specific items from each list be sent in electronic or hard copy form. As a follow-up to the gray literature solicitation letter, certain key organizations were contacted by phone to request their participation. Appendix B contains the gray literature solicitation letter and list of contacted utilities.

Project team members generated a list of vendors and private companies known to have experience in wastewater collection odor and corrosion. Each company on the list was contacted by telephone and asked to participate by submitting technical papers for inclusion in the project. Appendix C contains a list vendors contacted.

3.4 Previous WERF Literature Search Protocol

In order to consolidate information generated from a previous, related project and combine it with this project, literature from WERF's (2003) literature search and review was screened for relevance to this study; any papers from it that were related to collection systems were included here. The previous study included gray and published literature generated prior to 2000.

3.5 Search Results

References generated by the search processes were compiled and screened for inclusion. The first step in the screening process was to eliminate duplicate references, which composed perhaps 50% of the search hits. Next, references with titles obviously unrelated to the topic were eliminated. Abstracts of remaining references were reviewed for relevance. In general, all papers with direct relevance to wastewater collection odor and corrosion were included; also included were some papers with indirect relevance, including those related to the following:

- Treatment plants that had some discussion of processes upstream of the headworks
- Technologies for controlling odors generated in locations other than collection systems that might be applied to collection systems
- Collection system odor and corrosion as one among several topics

The published literature search generated 2,945 references, of which 263 were worth including. These papers were mostly conference proceedings, journal articles, and theses published in 2000 or later. Books were used to identify important primary sources and as references for individual authors, but books were not themselves included in the formal review process because books on this topic area usually provide a secondary compilation of work published in technical journals.

The gray literature search generated 52 papers from nine organizations. Of these, 33 contained information relevant to the project and were included in the system. From the previous WERF study, 317 papers were screened (published and gray), and 85 were included. The QA/QC process generated an additional 45 papers identified by project team members. The Googlescholar search produced 485 references, 14 of which were worth including and had not yet been identified. The total number of references screened was 3,843; 436 total papers were included. Nineteen of the papers turned out to be foreign language papers, and four could not be obtained. Thus the total number of papers submitted to the review process was 413. Table 3-2 summarizes these results.

Table 3-2. Literature Search Results Statistics.

Source	Number Submitted	Number Included
US Filter	3	3
Carus Chemical	5	0
Premier Chemical	15	6
LACSD	11	11
OCSD	4	4
City of Tacoma	2	2
Onodaga County	3	3
Ortec	6	3
Byogon	3	1
Previous WERF studies	317	85
Group 1 (Weiss) search	1,284	—
Group 2 (Basrai) search	672	—
Group 3 (Stuetz) search	634	—
Group 4 (Ward) search	355	—
Published literature subtotal	2,945	259
Identified by team members	45	45
Google Scholar search	485	14
Subtotal	3,842	436
Less those in a foreign language	—	19
Less those that could not be found	—	4
Total reviewed	**—**	**413**

Figure 3-1 summarizes the included literature in terms of percentage from each source.

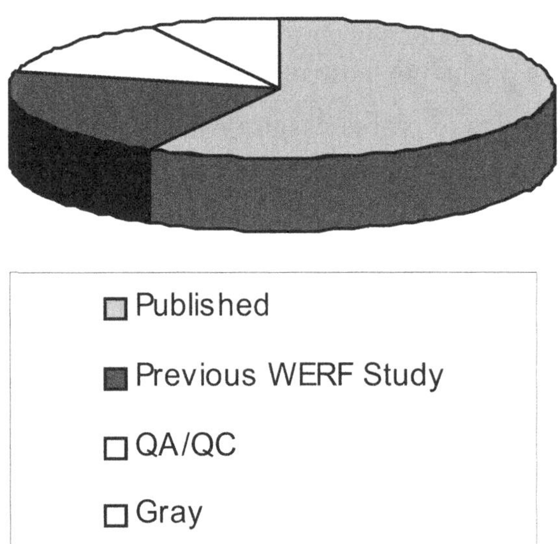

Figure 3-1. Distribution of Included Literature from Each Source.

Figure 3-2 summarizes the included literature in terms of percentage of each type.

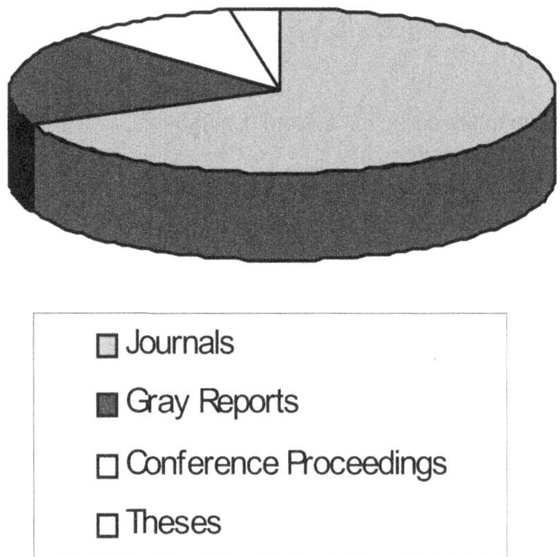

Figure 3-2. Distribution of Each Type of Included Literature.

All references were compiled and managed as Endnote files. Endnote is a software application designed to import, export, and manipulate bibliographic data. This format has features that may become useful later, in project phase II, during which a computer-based tool will be developed. Additionally, all references identified for inclusion were maintained in an Excel spreadsheet, which was posted on the project team website. Each reference identified for inclusion was assigned a unique identification number so that papers could be tracked without confusion due to nonunique bibliographic data.

3.6 Literature Acquisition and Management

A complete copy of each included reference was sought using one of five methods. All papers that could be obtained through library online journal subscriptions were obtained first. Papers not available through online journals were sought through other internet sources. Papers not available online were sought in hard copy form from several university libraries. All remaining references were requested through the interlibrary loan service of the University of Texas at Austin. Papers not available from a library were sought from project team members.

Each paper was converted to PDF and posted to the project website. A library containing paper copies of each paper was also created. Each paper was assigned, according to topic category, to one of nine primary authors. Table 3-3 lists each author along with his assigned topic categories.

Primary authors downloaded their assigned papers, reviewed them, and filled out a review form for each. Review forms were then copied into the Excel database. The review form contained fields for reference data, a summary topic statement, a quality rating, important conclusions, and identified research data gaps, and was structured to capture a paper's essential information and guide the author in composing the Phase I report. Information in the review form could also form the basis for a searchable database of information on each paper. Appendix D contains a copy of the review form.

Table 3-3. Phase I Report Primary Authors and Topic Categories.

Primary Author	Topic Category
Jay Witherspoon	Introduction and preliminary summary
James Joyce	Plain English guidance manual
	Sewer system overview and issues related to major collection system components
Richard Corsi &	Odor and corrosion compound generation and measurement
Richard Stuetz	Physical, chemical, and biological transformation processes in collection systems
Chris Quigley	Collection system emissions and emissions modeling
Steve Davidson	Collection System Ventilation approaches to reducing odors and corrosion in collection systems
Chris Easter & Matthew Ward	Liquid phase chemical treatment and liquid phase biological and enzyme treatment
Robert Bowker	Gas phase treatment
Philip Wolstenholme	Corrosion and corrosion protection
Chris Easter	Summary of research data gaps identified during the phase 1 literature search

Identifying topics worthy of further research was a primary purpose of the literature review process and will form the basis for later project phases.

3.7 Literature Search and Review Conclusions

Overall, the literature search and review process was systematic, thorough, and efficient. Peer-reviewed journal articles represented the bulk of new literature, indicating high-quality advances in the topic area. New academic theses were relatively few, indicating limited available funding in the topic area. The review process resulted in a comprehensive database of substantive commentary on each source. This will form the framework for a useful CD-ROM-based research tool, which can help to advance the quality of engineering practice in the area of wastewater collection system odor and corrosion.

CHAPTER 4.0

ODOR- AND CORROSION-CAUSING COMPOUNDS

4.1 Overview

Knowing the chemical composition of malodorous gas mixtures, the contribution of individual odorants to odor thresholds and intensity, and the mechanisms by which odorants are generated in sewers is key to developing optimal odor-monitoring and -control strategies. This chapter describes advances in identifying malodorous compounds emitted from sewers, biological mechanisms for odorous and corrosive chemical formation in sewers, and emerging measurement methods. Though over 60 papers were reviewed for this assessment, only a small fraction contributed new knowledge related to the subjects under consideration. Those papers are referenced where applicable throughout this chapter.

The knowledge base of odorous chemicals from sewers has not expanded appreciably during the past decade or more. Far more work related to odors from wastewater treatment processes appears to have been completed and will be referenced in this chapter where appropriate (for example, emerging odor measurement methods). The database related to chemicals that compose the mixtures of malodorous gases emitted from sewers is sparse and has limited advances toward understanding the specific chemical components that should be controlled to avoid odorous emissions from sewers. The knowledge base associated with corrosive gases is focused entirely on hydrogen sulfide. Hydrogen sulfide formation mechanisms, corrosion mechanisms, and prevention strategies are described in Chapter 9.0. A shorter summary of factors affecting hydrogen sulfide formation and release in sewers is provided in this chapter.

4.2 Malodorous Compounds Associated with Municipal Wastewater

Malodorous compounds are often characterized as small, hydrophobic, volatile molecules that contain one or two functional groups composed in part of oxygen, sulfur, or nitrogen. Several general classes of malodorous compounds that may be relevant to odors associated with sewers are listed in Table 4-1.

Table 4-1. Odorous Compounds Associated with Untreated Wastewater.

Class of Compound	Relevant Examples
Inorganic gases	Ammonia, hydrogen sulfide
Mercaptans	(Allyl, amyl, benzyl, crotyl, ethyl, methyl, propyl) mercaptans
Other organic S	Dimethyl sulfide, diphenyl sulfide, thiocresol, thiophenol
Amines	(Dibutyl, diisopropyl, dimethyl, ethyl, methyl, triethyl) amines
Diamines	Cadaverine (1,5-pentanediamine), putrescine (butanediamine)
Other organic N	Indole, pyridine, skatole
Volatile fatty acids	(Acetic, propionic, butyric, isovaleric) acids

The compounds listed in Table 4-1 have a wide range of characteristic odors. For example, hydrogen sulfide is characterized by rotten-egg odors, but several of the mercaptans cause odors characterized as decaying cabbage or vegetables, and other organic sulfur-containing compounds span a wide spectrum of odor characteristics, ranging from "skunk-like" to "garlic-like." Several amines are characterized by "fishy" odors. The two diamines are often characterized simply as "putrid." Two of the three other organic nitrogen-containing compounds (indole and skatole) are characterized as having strong fecal odors. Existing knowledge does not allow for accurately predicting the characteristic odors of various mixtures of these compounds. However, hydrogen sulfide, with its characteristic rotten-egg smell, is widely assumed to be the most prevalent malodorous and corrosive compound in the wastewater collection system. Additional details related to corrosion mechanisms are described in Section 4.4. The remainder of this section focuses on odorous gases.

Sato et al. (2001, 2002) completed two separate studies that taken together support the generally accepted belief that most odorous compounds emitted from sewers form during transport and are not appreciably discharged directly to municipal sewer systems. Sato et al. (2002) identified and quantified odorous gases associated with freshly excreted human feces. Toilet headspace samples were collected immediately after human defecation. The concentrations of malodorous compounds were generally low: 5-26 ppb for hydrogen sulfide, 2-15 ppb for methyl mercaptan, less than 100 ppb for ammonia, and approximately 10 ppb for volatile fatty acids (VFAs). Interestingly, concentrations of VFAs, particularly acetic acid, were five orders of magnitude higher for subjects with diarrhea than for normally healthy subjects. This implies that direct discharges of VFAs may be significant during flu season, but additional research would be needed to quantify the contribution of such discharges.

Sato et al. (2001) also collected wastewater (human waste) samples from a storage tank at a community wastewater treatment plant (it was not stated but is assumed that the wastewater was sampled prior to treatment and is representative of end-of-sewer samples). Liquid samples were transferred to the gas phase through purging with helium and were analyzed by gas chromatography with mass spectrometry (GC/MS) (nonsulfur compounds) or gas chromatography with flame photometric detection (GC/FPD) (sulfur species). Approximately 90% of the malodorous compounds on a gas phase molar basis were C2 to C5 VFAs. Acetic acid was the dominant VFA (up to 120 ppm), followed by propionic acid (up to 27 ppm). Sulfur-containing compounds were dominated by hydrogen sulfide (up to 50 ppm) and methyl mercaptan (up to 1.1 ppm). Nitrogen-containing compounds were dominated by ammonia (up to 34 ppm) and trimethylamine (up to1.2 ppm). Pyridine, pyrrole, indole, and skatole were also detected. Dichlorobenzene, a common deodorizer used in household toilets, was also detected.

It is clear that most malodorous compounds associated with municipal sewers actually form during transport. Biological transformations that lead to the formation of VFAs and other odorants associated with sewers are described in Section 4.3.

4.3 Odorous Gas Formation Mechanisms

Sewer collection systems offer a variety of environments for microorganisms to perform a wide range of microbial transformation processes. This variability is induced by the large flow fluctuations typically observed in most sewer networks, causing changes to the dissolved oxygen (DO) availability in the collection system and promoting erosion or deposition of sewer sediments and erosion and growth of biofilms. Aerobic conditions are often observed in partially filled gravity sewers, whereas anaerobic conditions occur generally in pressure sewers and full-flowing gravity sewers (Hvitved-Jacobsen and Volertsen, 2001) as well as inside some sewer biofilms and sediments. Because of the low level of nitrate concentrations in most wastewaters, anoxic conditions are generally not significant in sewers (Hvitved-Jacobsen, 2002).

Aerobic and anoxic microbial degradation of wastewater do not generally contribute significantly to the generation of odorous compounds in collection systems (Hvitved-Jacobsen and Volertsen, 2001). As such, this section focuses on anaerobic heterotrophic microbial processes that are involved in the generation of the odorous substances typically associated with sewer odor complaints.

Figure 4-1 shows the main anaerobic mechanisms related to carbon transformation in sewers: anaerobic hydrolysis, fermentation, methanogenesis, and sulfate reduction.

Anaerobic hydrolysis is an important precursor to the other anaerobic processes involving the transformation of large, complex organic molecules into smaller, simpler molecules that can be directly utilized by other microorganisms as a substrate. Hydrolysis directly provides the substrate for fermentation and sulfate reduction and indirectly provides substrate for both methanogenesis and sulfate reduction (via fermentation as an intermediate process). The organic products of anaerobic hydrolysis generally do not contribute directly to odor emissions from sewers, with the exception of the hydrolysis of organic sulfur-containing compounds by bacteria such as *Proteus spp.*, *Bacteroides spp.*, and *Clostridium spp.*, resulting in the production of hydrogen sulfide, organic sulfides, and disulfides (Hvitved-Jacobsen and Volertsen, 2001). However, in sewer systems without significant accumulation of sediments, fermentation and sulfate reduction are the dominant odor production pathways (Hvitved-Jacobsen, 2002).

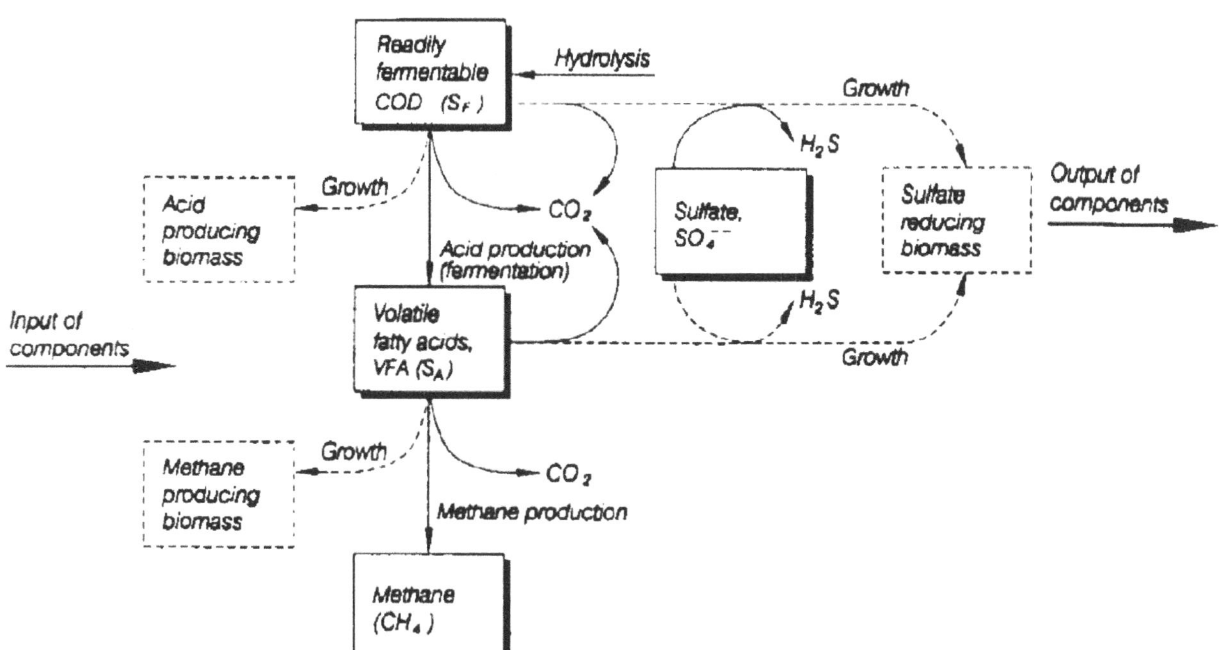

Figure 4-1. Anaerobic Carbon Transformations in Sewer Systems. (Reproduced with permission. From Hvitved-Jacobsen, 2002)

Microbial fermentation involves the transformation of organic products (that is, readily biodegradable substrate) from anaerobic hydrolysis into fermentation products such as low-molecular-weight organic products and carbon dioxide. These organic products generally include a wide range of VFAs and volatile organic compounds (VOCs), which can contribute to the overall odor emissions potential from a sewer system (Hvitved-Jacobsen, 2002).

Fermentation products can be used by both methanogenic and sulfate-reducing bacteria (SRB) as a substrate (Hvitved-Jacobsen and Volertsen, 2001); however, methanogens compete with SRB within biofilms and sediments. Typically, methanogenesis occurs only in areas where sulfate does not penetrate into the biofilms and sediments, and it is therefore regarded as a minor anaerobic process leading to odor formation (Hvitved-Jacobsen, 2002); sulfate reduction, on the other hand, contributes significantly to the production of hydrogen sulfide and is one of the key microbial processes involved in the generations of odorous and corrosive compounds in sewers.

4.3.1 Fermentation

Fermentation processes can occur in all of the major parts of the sewer system (in sediment, biofilms, and bulk wastewater), provided that a suitable anaerobic environment is present (Hvitved-Jacobsen, 2002). Short-chain organic molecules are produced during fermentation through oxidation and reduction reactions that use organic matter as both the electron donor and acceptor. In general, VFAs are generated through the fermentation of carbohydrates (for example, starch), whereas other sulfur-containing compounds, such as mercaptans, are generated through the fermentation of proteins (Hvitved-Jacobsen, 2002).

The specific fermentative pathways and products will be highly dependent on the substrates and microorganisms present (Hvitved-Jacobsen, 2002), but they can include many of the typical odorous substances—ammonia, amines, alcohols, aldehydes, ketones, VFAs, and organic sulfides such as ethyl mercaptan, dimethyl sulfide, methyl disulfide, and hydrogen sulfide (Hvitved-Jacobsen and Volertsen, 2001). As well as producing odorous compounds during fermentation, VFA creation can lead to a reduction in the pH of the wastewater, which in turn promotes the further release of organic sulfides and hydrogen sulfide into the sewer atmosphere (Hvitved-Jacobsen and Volertsen, 2001), thereby potentially exacerbating the odor problems caused by SRB.

4.3.2 Sulfate Reduction

SRB are anaerobes and as such are relatively slow growing owing to the low energy yield of the sulfur reduction process. SRB will typically be washed out of the bulk wastewater; thus the majority of the sulfate reduction process occurs in sewer sediments and biofilms—sewer slime—where the bacteria are protected from the shear forces associated with the flow (Hvitved-Jacobsen, 2002). Some sulfate reduction can occur in the bulk wastewater flow because of biofilm entrainment; however, this is generally negligible compared to the processes occurring in the sediment and biofilms (Hvitved-Jacobsen, 2002). Generally, sulfate reduction in biofilms is the dominant sulfide formation mechanism in pressure sewers, whereas sulfate reduction in biofilms and sediments is important in gravity sewers (Hvitved-Jacobsen, 2002).

Some common species of SRB are *Desulfovibrio spp.* and *Desulfotamaculum spp.* (Hvitved-Jacobsen, 2002). Ito et al. (2002) studied the speciation and metabolism of SRB present in an oxic-anoxic sewer biofilm using fluorescent in situ hybridization combined with microautoradiography. SRB were observed to be a small fraction (4.8%) of the total cells present in the biofilm, with 70% of these identified as belonging to the Desulfovibionaceae family, and the remaining 30% to the Desulfobacteriaceae family. Further observation showed that less than

50% of the total SRB from the Desulfovibionaceae family was identified as *Desulfobulbus spp.* and *Desulfovibrio spp.* Less than 7% of the Desulfobacteriaceae family species were identified as *Desulfonema spp.*, suggesting that many species of SRB could not be identified with the probes used in the study. It was also demonstrated that SRB are able to adapt to rapidly changing environments (in terms of oxygen availability), with a percentage of the Desulfobulbus cells having the ability to utilize propionate with oxygen or nitrate as the electron acceptor (9% and 27%, respectively), and more than 40% of Desulfobulbus cells being able to utilize acetate under anoxic conditions (Ito et al., 2002).

The mechanism of sulfate reduction involves sulfate being reduced to hydrogen sulfide directly using organic carbon as a terminal electron donor according to the following generalized reaction pathway (Hvitved-Jacobsen, 2002):

$$SO_4^{2-} + \text{organic carbon} \rightarrow HCO_3^- + H_2S \qquad (4.3.2\text{-}1)$$

Depending on the substrate and type of bacteria present, low-molecular-weight organics such as acetate can be formed instead of bicarbonate (or carbon dioxide) (Hvitved-Jacobsen, 2002). The concentrations and availability of both sulfate and suitable organic carbon substrate are key aspects that control the type and kinetics of sulfide formation. The hydrogen sulfide formed via reaction 4.3.2-1 is not only one of the dominant odorous chemicals in sewers but also the primary source of sewer corrosion. The role of hydrogen sulfide as a source of sewer corrosion is summarized in Section 4.4, and additional details are provided in Chapter 9.0.

The amount of sulfate present in a collection system (and hence available for reduction and sulfide production) is dependent on the sulfate content of the raw wastewater (inorganic sulfates and sulfonates). This can also be influenced by groundwater hardness, potable-water treatment methods, and industrial sources (Hvitved-Jacobsen and Volertsen, 2001). When dealing with industrial inputs to collection systems, it is important to consider that other oxidized sulfur compounds, such as thiosulfate and sulfite, may be used as sulfur sources by SRB (Nielsen, 1991). Typically, raw municipal wastewater has sulfate concentrations greater than 5 to 15 g S/m^3, which for thin biofilms will not be the limiting condition for sulfide formation (Hvitved-Jacobsen, 2002). However, for thicker biofilms or biofilm-covered sediments, the increased resistance to sulfate transfer can result in sulfate-limited conditions that may enable methanogenesis to occur, as noted by Vollertsen and Hvitved-Jacobsen (2000), who observed methane-containing gas cavities inside incubated beds of sewer sediments.

SRB use short-chain organic molecules (typically fermentation products such as alcohols, lactate, pyruvate, and some aromatics) as substrates. The biodegradability and suitability of the substrate is an important parameter affecting the sulfate reduction rate; for example, Hvitved-Jacobsen (2002) notes that enhanced sulfate reduction could occur in sewer systems accepting food-processing waste (which provides a more concentrated supply of readily biodegradable organic matter) or in sewer systems in areas where the organic substrate concentrations have been increased because of water shortages or water reuse. The form of exposure to organic substrates can also affect sulfate reduction; experiments conducted by Norsker et al. (1994) with laboratory-scale biofilm reactors indicated that pulse dosing of organic substrates (similar to concentration variations often found in sewers) increases the organic substrate availability to SRB and hence the sulfate reduction kinetics.

In addition to the sulfate and organic carbon concentrations, temperature and DO concentration impact the formation of sulfides. Wastewater pH is generally not a significant influence on sulfide production because SRB can exist between pH 5.5 and pH 9 and are not inhibited below pH 10 (Hvitved-Jacobsen, 2002). However, it is important to note that pH does have an important influence on the sulfide species equilibrium in the bulk wastewater and will affect the release of compounds into the sewer atmosphere (Hvitved-Jacobsen, 2002). Temperature can have a strong effect on sulfate reduction kinetics, with a temperature dependency coefficient of 1.13 being reported, although this coefficient is reduced (on the order of 1.03) for biofilms or sediments where substrate diffusion is the limiting condition (Hvitved-Jacobsen, 2002). This temperature dependency is applied as follows:

$$R_{S,T} = R_{S.20} = \alpha^{(T - 20)} \qquad (4.3.2\text{-}2)$$

Where

$R_{S,T}$ = rate of sulfide production at temperature T (g/m^2/hr)

$R_{S.20}$ = rate of sulfide production at 20°C (g/m^2/hr)

α = temperature dependency coefficient, e.g. 1.13 (unitless)

T = wastewater temperature (°C)

Also, adaptation in lower- or higher-temperature climates may reduce the apparent temperature dependency (Hvitved-Jacobsen, 2002).

In sewer systems, oxygen consumption can occur rapidly. Hvitved-Jacobsen (2002) suggests that if the bulk wastewater DO concentration is maintained above 0.2 to 0.5 g O_2/m^3, sulfide emissions to the sewer atmosphere should not be significant. The objective is not to avoid sulfide formation (in the sewer biofilm) but to keep the bulk sulfide at a low-concentration level in order to control odor problems. Norsker et al. (1994) demonstrated that the variations in DO conditions under which the biofilms are established can also have a strong influence on the sulfate reduction potential. Although biofilms grown at all oxygen levels had the potential to produce sulfide, those grown at very low DO concentrations had the greatest sulfide production potential (Norsker et al., 1994).

4.4 Hydrogen Sulfide as a Corrosive Gas

Hydrogen sulfide causes direct corrosion of metals such as iron, zinc, copper, lead, and cadmium. More importantly, hydrogen sulfide forms sulfuric acid, which can cause major damage to wastewater facilities, including the crown of concrete sewer pipes.

Corrosion of wastewater pipelines is caused primarily by microbially induced corrosion (MIC). This is a complex process involving many active organisms that are not fully understood. There are two distinct phases of MIC: The first is the conversion of ubiquitous sulfate in the wastewater to sulfide, some of which is released as gaseous hydrogen sulfide; the second is the conversion of hydrogen sulfide to sulfuric acid, which attacks susceptible pipeline materials.

As described in Section 4.3, the bacteria producing sulfide are generally considered strict anaerobes and, consequently, live beneath the water surface in gravity sewers and on the pipe walls in force mains. The bacteria may also thrive in sludge and grit deposits found along the bottom of sewer pipes. In order to produce sulfide compounds, the bacteria require a source of sulfur and a food supply. Sulfate, generally abundant in wastewater, is the common sulfur source. However, other forms of sulfur such as organic sulfur from animal wastes can also be reduced to sulfide. Dissolved organic material prevalent in the wastewater provides an ample food supply for bacteria to flourish. The sulfate reaction in a wastewater collection system can be defined in accordance with Equation 4.3.2-1. The overall sulfide and sulfuric acid formation mechanisms are illustrated in Figure 4-2.

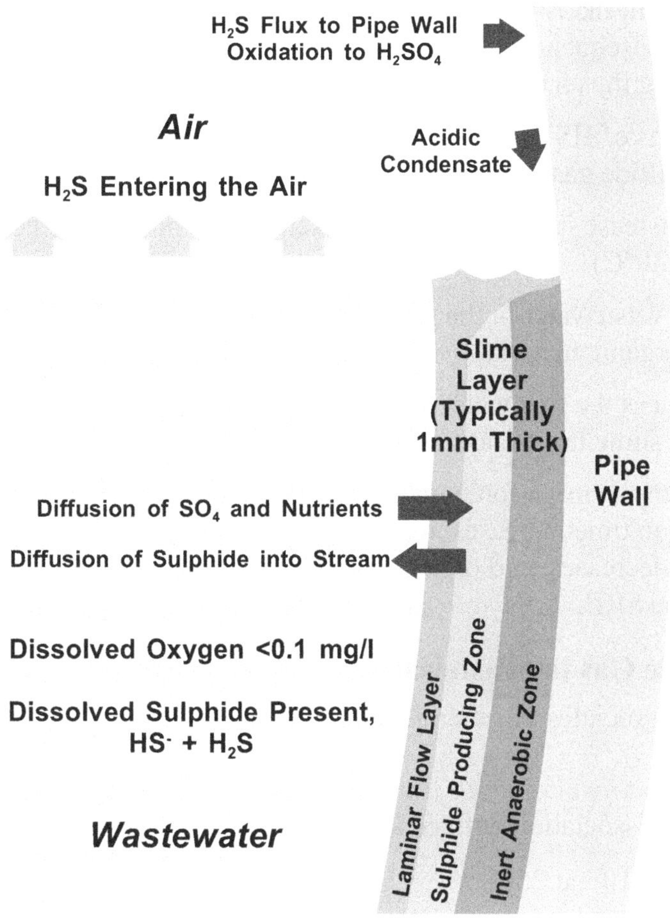

Figure 4-2. Overall Corrosion Mechanism in Wastewater Pipelines. (With permission from ASCE, 1989)

Sulfate and nutrients diffuse into a biofilm that grows on the submerged wetted walls of sewer pipes. There, the sulfate is reduced to sulfide, primarily through the action of anaerobic SRB. The hydrogen sulfide then diffuses out of the biofilm and into the wastewater, where it dissociates primarily between hydrogen sulfide (H_2S) and bi-sulfide (HS^-). The uncharged hydrogen sulfide can be released to the overlying sewer atmosphere, absorbed to water films on the sewer pipe, and oxidized to sulfuric acid.

4.4.1 Rate of Sulfide Production

The rate of sulfide production is dependent upon environmental conditions in the slime layer. The following wastewater conditions are the most critical parameters impacting the rate of sulfide production:

- ***Concentration of organic material and nutrients.*** These materials diffuse into the slime layer and are consumed by the SRB.
- ***Sulfate concentration.*** Sulfate and organic matter will be used by the SRB in the ratio of approximately 2:1, depending upon the relative concentrations of each.

- ***Dissolved oxygen.*** DO is critical in determining whether anaerobic or aerobic bacteria will dominate the breakdown of organic material in the wastewater. If the DO concentration is in excess of 1.0 mg/L, aerobic bacteria will likely dominate the activity, particularly on the outer layers of an attached biofilm. Consequently, increased DO will reduce the production of sulfide by limiting the food reaching the anaerobic bacteria.
- ***pH.*** The pH determines the proportions of HS^- and H_2S found in the wastewater and directly influences the amount of hydrogen sulfide gas available for release into the atmosphere.
- ***Temperature.*** Each degree Celsius increase in temperature represents a 7%increase in the biological activity of the SRB (up to 30°C).
- ***Wastewater velocity.*** Velocity of the wastewater in the sewer influences the thickness of the slime layer and deposition rates for organic material.
- ***Surface area.*** The flow depth influences the free water surface and determines the submerged pipe surface in which the slime layer may form.
- ***Detention time.*** This is particularly important in force mains, submerged sewers, and sewers with little DO present. As the detention time increases, more oxygen is consumed, oxidation-reduction potential (ORP) decreases, and organic material becomes more solubilized, a condition that favors the SRB.

4.4.2 Solubility of Hydrogen Sulfide Gas in Wastewater

Dissolved hydrogen sulfide gas dissociates in water as follows:

$$H_2S \Leftrightarrow HS^- + H^+ \qquad (4.4.2\text{-}1)$$

The equilibrium constant K_{H2S} is for this dissociation defined as:

$$[HS^-] \times [H^+]/[H_2S] = K_{H2S} = 1.12 \times 10^{-7} \text{ at } 25°C \qquad (4.4.2\text{-}2)$$

where $[HS^-]$, $[H^+]$, and $[H_2S]$ denote the molar concentrations of the respective constituents. Consequently, if hydrogen ions are present in the wastewater, the concentration of dissolved hydrogen sulfide gas must increase as bacteria produce sulfides so the chemical equilibrium of the above dissociation reaction is maintained. The production of hydrogen sulfide is highly sensitive to the concentration of hydrogen ions (i.e., pH). A reduction of the wastewater pH indicates increased hydrogen ions in the wastewater and causes an increase in dissolved hydrogen sulfide gas concentration. Conversely, increasing the pH causes a reduction of the dissolved hydrogen sulfide gas concentration.

A second dissociation from HS^- to S^{2-} can occur; however, generally, S^{2-} only exists at pH levels greater than 12, which is well above the normal wastewater pH range.

Because hydrogen sulfide gas is relatively insoluble in water, its solubility can be described by Henry's Law as follows:

$$X_{H2S} = K_{H2S}P_{H2S} \qquad (4.4.2\text{-}3)$$

where X_{H2S} = mole fraction of dissolved hydrogen sulfide at equilibrium in the liquid phase (moles H_2S/moles liquid), K_{H2S} = Henry's law constant for hydrogen sulfide gas at a given temperature (moles/liter H_2S/atm H_2S), and P^*_{H2S} = partial pressure of hydrogen sulfide gas at equilibrium in the atmosphere (atm).

Under most conditions in the sewer, the actual partial pressure of hydrogen sulfide in the sewer headspace (P_{H2S}) is low relative to the saturation partial pressure (P^{*}_{H2S}). Thus, there is a hydrogen sulfide gradient from the surface of wastewater to the overlying sewer atmospheres and consequential driving force for emissions of H_2S from wastewater to sewer air. Bacteria continuously produce sulfides that are converted to hydrogen sulfide, which is released to the atmosphere, where it can cause corrosion and odor and safety problems.

4.4.3 Corrosion of Sewer Walls

Once sulfides are produced in the wastewater as the result of sulfate reduction, hydrogen sulfide gas will be released into the atmosphere. The hydrogen sulfide gas is oxidized on the pipe surface above the water line as described by the following equation:

$$H_2S + O_2 \xrightarrow{\text{bacteria}} H_2SO_4 \qquad (4.4.3\text{-}1)$$

Oxidizing bacteria, such as *Thiobacillus concretivorous*, *Thiobacillus neapolitanus*, *Thiobacillus ferro-oxidans*, or *Thiobacillus thioxidans*, are aerobic and thrive in low-pH environments. To grow, they require a source of sulfur (H_2S), a moist surface (the pipe walls above the water surface), and a carbon source (CO_2). The surface pH of new concrete pipe is generally in the range of 11 to 13. Cement contains calcium hydroxide which neutralizes the acids and inhibits formation of oxidizing bacteria when the concrete is new. However, as the pipe ages, the neutralizing capacity of the pipe is consumed, the surface pH drops, and the sulfuric acid-producing bacteria become dominant. In active corrosion areas, the surface pH can drop below 2.0. As sulfides are formed and sulfuric acid is produced, calcium hydroxide is converted to calcium carbonate (gypsum), which is easily eroded by wastewater.

4.5 Odor-Sampling and Analysis Methods

The previous subsections involved the nature of malodorous and corrosive compounds associated with sewers and the biological transformation mechanisms that can lead to their formation. Methods for sampling and analyzing malodorous compounds and mixtures are considered in this subsection.

Odor-sampling methods can be divided into three broad classifications: 1) human sensory evaluations, that is, perceptions of odor thresholds, concentrations, and pleasantness by odor panelists; 2) measurement of chemical-specific surrogates for odor, for example, hydrogen sulfide; and 3) measurements of nonspecific odor fingerprints or profiles, for example, "electronic nose" systems. Each of these approaches is described below.

4.5.1 Sensory Evaluations

Conventional measurements of odor involve human sensory evaluations (Figure 4-3). An olfactometer (Figure 4-4) is typically employed to quantify odor detectability by using sequentially decreasing dilutions-to-threshold (D/T) values and presenting the results as odor units (OUs) or threshold odor numbers (TONs). Odor intensities may also be measured by subjective comparisons to a reference chemical such as n-butanol. While these methods are frequently employed to assess odors associated with wastewater, there have been few studies reported for sewers. Furthermore, sensory evaluation methods in the present have been limited, owing to uncertainties related to large differences in sensitivities to odors among odor panelists, and the aging of samples between sample collection and employment of an odor panel

(McGinley and McGinley, 1997; Walsh, 1996); however, more-recent standardizations of panelists' selection and olfactory-operating protocols using the European CEN standard for dilution olfactometry have significantly improved the repeatability of olfactory determinations. Figure 4-5 shows the ratio of maximum to minimum odor concentrations observed by members of the same odor panel for the same gas mixture. The majority of ratios are between 1 and 3 with, as expected, greater variations at low odor concentrations.

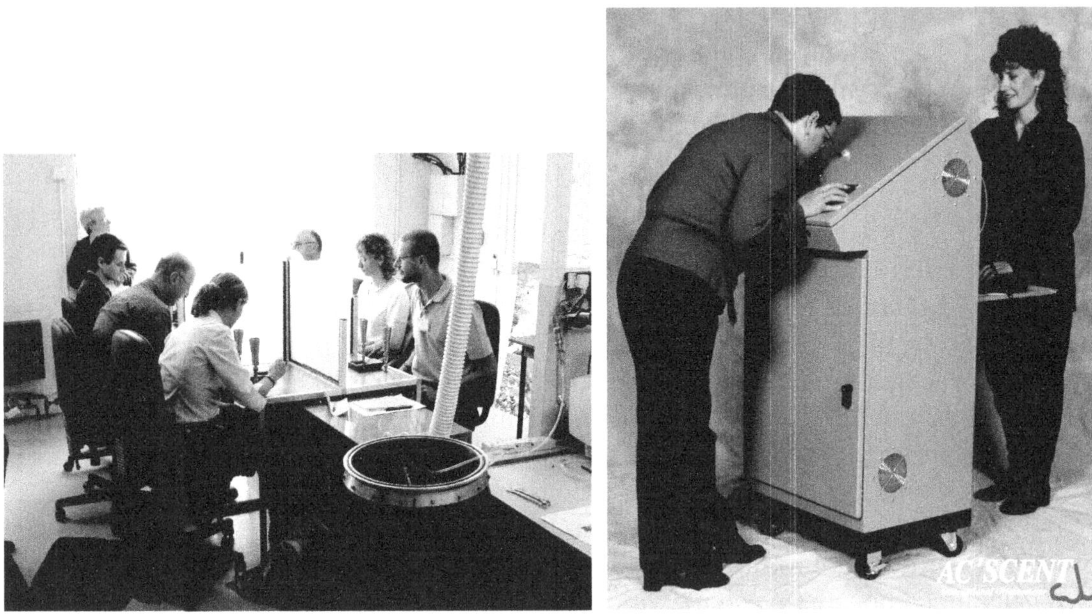

(a) (b)

Figure 4-3. Olfactory Evaluation of Odor Samples by Panelists (a) Sitting in the Same Room and (b) Using an Olfactometer While Isolated from Other Panelists (Panel a courtesy of Silsoe Odours Ltd, Bedford, UK; panel b courtesy of St. Croix Sensory)

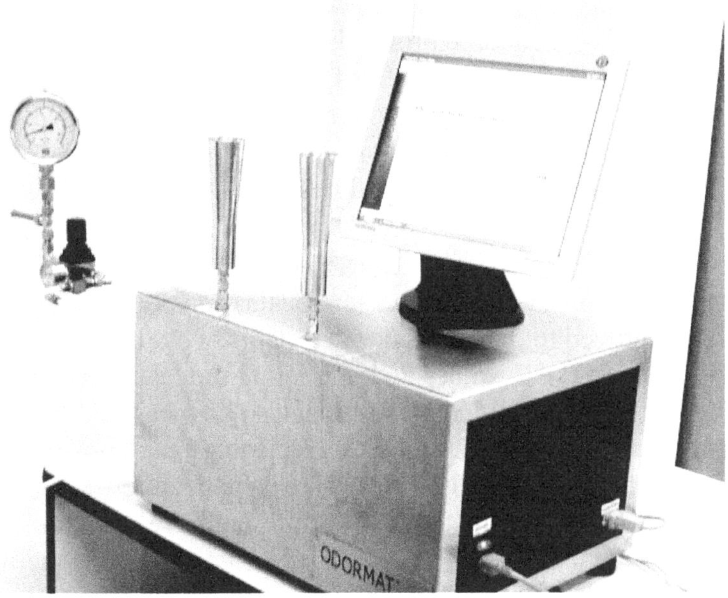

Figure 4-4. Example of Olfactometer with Two Sniffing Ports. (Courtesy of Aromatrix, Singapore)

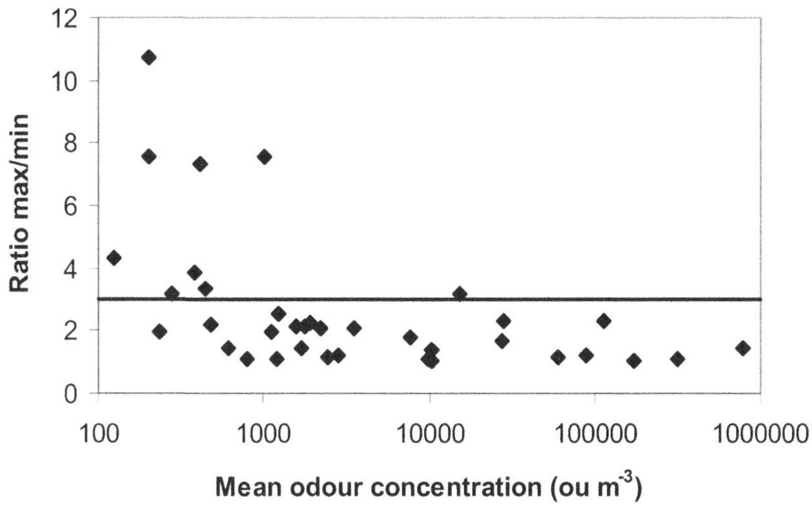

Figure 4-5. Repeatability of Paired Sewage Odor Samples. (After Gostelow and Parsons, 2000)

There remains significant uncertainty regarding the effects of mixtures of wastewater odorants on the sensory evaluation of related odors, and few papers on this subject have been published over the past decade. Laing et al. (1994) completed a systematic study to assess the effects on human sensory evaluations of chemical mixtures of odors relevant to wastewater treatment facilities. They studied hydrogen sulfide, isovaleric acid, butanethiol, and skatole as well as binary, trinary, and quaternary mixtures of these odorants. The perceived odor intensities of mixtures were greater than any individual components but less than the sum of intensities of individual components. As the number of components in a mixture increased, there was greater suppression of the effects of individual components. However, the intensity of the most-dominant component was observed to provide a good approximation of the intensity of the mixture. The least frequently suppressed component was observed to be hydrogen sulfide; the most frequently suppressed components, isovaleric acid, and skatole. Importantly, the "unpleasantness" of mixtures was usually greater than individual components of the mixture, and as such, methods or models based on single odorants may significantly underestimate the number of odor complaints.

There may be value in characterizing different wastewaters in terms of their potential to lead to malodorous gases. To measure the odor emission of a liquid wastewater sample, a methodology is needed that relates the emission of odorants to the volume of the respective liquid rather than to the surface area as determined by conventional wind tunnel and emission hood measurements. Two methodologies have been developed to evaluate this potential effect: odor potential (OP) (Hobson, 1995) and odor emission capacity (OEC) (Frechen and Köster, 1998). Both methods involve measuring the odor concentration of air bubbled through a wastewater sample under standardized conditions (Figure 4-6). The air sample is typically collected in an inert container (for example, Tedlar bag, polished stainless steel canister) for conventional sensory evaluations—for example, dilution olfactometry analysis to determine TONs—or determination of specific odorants, such as H_2S. Having applied the OP concept to a range of municipal wastewaters, Hobson (1995) reported TONs ranging over four orders of magnitude for untreated wastewater arriving at a wastewater treatment plant. Such a large range of OP indicates substantial variations in either the septicity or chemical makeup of raw wastewater and clearly shows that there is no such thing as "typical" untreated wastewater. It is important to recognize that the OP method does not account for the potential for a wastewater to form odorants through chemical or biological transformations; that is, it accounts for only a current wastewater state.

Figure 4-6. Odor Emission Capacity Test Reactors.

4.5.2 Studies Involving Hydrogen Sulfide and Other Sulfur Species

Field methods to assess odors associated with wastewater have relied heavily on the measurement of individual odorants (such as hydrogen sulfide concentrations at the fence line of a wastewater treatment plant) using electrochemical or gold film sensors; see Figure 4-7. (Gold film sensors, though used specifically for H_2S measurement, are sensitive to the presence of other sulfides). However, there have been few advances in the use of hydrogen sulfide measurements as a means of investigating odor problems, particularly as related to sewers, during the past decade. Several examples are described below; they relate to wastewater treatment plants and other sources, but they may have some application to sewers. As described in Section 4.4, the release of hydrogen sulfide from wastewater is also a major source of corrosion in municipal sewers. As such, methods for measuring hydrogen sulfide gas are also relevant to detecting environments conducive to sewer pipe corrosion.

Figure 4-7. Hydrogen Sulfide Monitor (Jerome 631-x). (Courtesy of Arizona Instruments)

Witherspoon and Barnes (2004) investigated the use of hydrogen sulfide and OU source measurements coupled with a dispersion model to predict odors at the fence line of a municipal wastewater treatment plant. Ambient measurements of hydrogen sulfide and odor D/T values were made (using a field olfactometer) at fence line. Application of a dispersion model significantly overestimated fence line hydrogen sulfide concentrations, but the average ratio of predicted to measured odor concentrations across 20 locations was reasonably good (0.79). Although the specific application involved wastewater treatment processes, the results suggest a possible application to sewers, particularly as related to determination of specific emission sources that might contribute to community odor complaints.

Gostelow and Parsons (2000) gathered previously reported data from odor surveys at 17 wastewater treatment plants to assess the possibility of linking analytical measurements of hydrogen sulfide and sensory evaluations of odor quantified as number of OUs per square meter. All data were obtained for covered treatment processes, both before and after odor control. Correlations between odor concentrations and hydrogen sulfide suggest a power law relationship of the following form:

$$C_{OU} = mC^{n}_{H2S} \qquad (4.5.2\text{-}1)$$

where C_{OU} is the odor concentration as measured by sensory evaluation (OU/m^3), C_{H2S} is the hydrogen sulfide concentration (ppm), and m and n are empirical constants. For preliminary treatment prior to odor control (most representative of untreated wastewater in sewers), m was observed to equal 52,555, and n was reported as 0.62. The value of R^2 was only 0.45, indicating that only 45% of the variation in odor concentration was explained by variations in hydrogen sulfide. The remainder (55%) was explained by other odorants and/or measurement errors. For comparison, values of m and n for aeration tanks were 14,555 and -0.12 ($R^2 = 0.07$), respectively. Values of m and n for sludge storage and handling were 38,902 and 0.64 ($R^2 = 0.69$), respectively. These results are consistent with work by Hobson (1995), who noted that the TONs associated with biogenic odors from wastewater are frequently 5-10 times greater than that expected from hydrogen sulfide concentrations. As such, other chemicals clearly play an important role in defining human perception of wastewater-related odors.

Finally, Sye and Chen (2000) described an analytical method to determine a range of sulfur species at parts-per-billion levels in odorous gas mixtures. Sulfur species in air were collected onto an adsorbent (Tenax-GR) with subsequent analysis by thermal desorption, separation by gas chromatography using a fused silica porous-layer open tubular capillary column, and detection by flameless sulfur chemiluminescence (FSC). The use of FSC for detection was concluded to be an excellent alternative to the more common flame photometric detection. Linear responses were observed for sulfur species, including hydrogen sulfide, sulfur dioxide, dimethyl sulfide, dipropyl sulfide, and C1 to C7 speciated thiols. The method was tested for air samples collected above an aerated treatment tank at a petrochemical plant and a hoggery.

4.5.3 Solid Phase Microextraction

Solid phase microextraction (SPME) was first introduced in the early 1990s as a method for concentrating trace chemicals prior to their analysis by chromatographic techniques. It holds promise for the capture of gas phase odorous compounds associated with municipal wastewater with subsequent resolution and quantification at levels of parts per trillion by volume to parts per billion by volume.

SPME involves a sorptive medium (liquid or solid) that is applied as a thin layer over a fused silica fiber. Common coatings include polydimethylsiloxane (PDMS), polyacrylate, and Carboxen. When the coated fiber is placed in a fluid such as air or water, chemicals in the fluid are absorbed into the coating and concentrated prior to analysis by gas chromatography with any of a number of detectors. There are two primary modes associated with the use of SPME fibers: 1) direct sample collection, during which the SPME fiber is exposed directly to the medium of interest (for example, in an air or wastewater stream or sample), and 2) headspace SPME (HS-SPME), in which an SPME fiber is exposed to the gaseous headspace above a liquid or solid sample matrix.

The benefits of SPME for sample collection and preconcentration are numerous and include the ability to

- Concentrate a wide range of analytes
- Use coated fibers in air, water, or wastewater
- Concentrate and analyze trace contaminants that might have otherwise gone undetected with conventional field monitoring or whole-air samples
- Directly desorb chemicals from an SPME fiber with zero-path injection onto a gas chromatographic column
- Analyze without the use of solvents

The major drawback to SPME is the need to establish chemical-specific partition coefficients between the medium of interest and the fiber coating, and to account for variations in medium temperature, relative humidity, and fluid mechanic conditions around the fiber, each of which might be important variables if samples were collected in the actual headspace of sewers.

The use of SPME for air sampling emerged in the mid- to late 1990s and has been described in numerous papers (for example, Martos and Pawliszyn, 1997). Applications related to sampling sewer gas are sparse. However, several recent studies have focused on the use of SPME for identifying and quantifying odorous gas emissions from various wastewater processes, suggesting its potential use for characterizing low concentrations of odorous chemicals in untreated wastewater, in the headspace of sewers, or downwind of points where sewer odor is released. The latter may be the most relevant application, given that SPME is a technology that is particularly valuable for measuring concentrations of compounds in air at levels of parts per thousand to parts per billion, and such levels are not likely to cause odor complaints when immediately released from sewers. However, higher concentrations can be diluted to parts-per-thousand to parts-per-billion levels downwind of a release point, and being able to effectively identify and quantify an odor fingerprint (chemical composition) downwind of an odor release might be useful for back-calculating the nature and magnitude of release from the source. Several relevant studies are described below.

Kim et al. (2002) used SPME to concentrate odorous chemicals from the headspace of wastewater samples collected from several processes (wastewater influent, thickened sludge, dewatered sludge, and biosolids) at a municipal wastewater treatment plant. A mixed Carboxen-PDMS-coated fiber was used to capture trimethylamine, carbon disulfide, dimethylsulfide, and dimethyldisulfide. A polyacrylate-coated fiber was used to capture propionic and butyric acids. Samples were analyzed by either gas chromatography with mass spectrometry (GC/MS) or gas chromatography with flame ionization detection (GC/FID) (for the carboxylic acids). It was concluded that SPME coupled with GC/FID or GC/MS can be effective for capturing the spectrum of chemicals listed above at low levels of parts per billion. Kim et al. (2002) observed the major reduced sulfur compound in wastewater influent (sewer effluent) to be carbon disulfide.

Huang et al. (2004) employed headspace solid phase microextraction (HS-SPME) coupled with either gas chromatography with nitrogen-phosphorus detection (HS-SPME-GC/NPD) or gas chromatography with mass spectrometry (HS-SPME-GC/MS) to analyze speciated odor components associated with a subsurface-constructed wetland for wastewater treatment. They were able to resolve and quantify a range of VFAs (C2 to C7), volatile alkyl sulfides (dimethylsulfide and dimethyldisulfide), and organic amines (methyl amine, dimethylamine, trimethylamine ethylamine) in wastewater from levels of parts per trillion by volume to parts per billion by volume using PDMS as the SPME coating. It was concluded that for the system tested, removal of propionic acid and isovaleric acid is particularly critical for deodorizing wastewater.

Turkmen et al. (2004) optimized an SPME-GC/MS method for analyzing selected odors known to be generated by digested sludge. The method involved a Carboxen-PDMS-coated fiber and was used to quantify parts-per-billion levels of sulfurous odor vapors such as dimethylsulfide, dimethyldisulfide, methyl mercaptan, hydrogen sulfide, and carbon disulfide. The authors concluded that dynamic measurements (direct air samples in the field) are suspect unless the fluid mechanic conditions of the calibration standard and field sample are identical. This problem does not exist with static headspace samples, for example, above a wastewater sample. The authors also noted the need to conduct additional research to ascertain possible errors associated with multisorbate competition and to lower detection limits for a range of malodorous chemicals to levels that can be detected by the human nose.

James and Stack (1997) employed both polyacrylate- and PDMS-coated SPME fibers for direct sampling of wastewater and the headspace above wastewater samples. Municipal wastewater, pharmaceutical, and petrochemical discharges were studied. Only direct immersion was used with municipal wastewater. Method detection limits for concentration in wastewater varied between 10 and 170 μg/L, depending on the specific chemical. Thirteen VOCs were identified and quantified for municipal wastewater, including a range of saturated straight-chain VFAs (C2, C3, C4, C6, C12, and C14) and cyclohexane carboxylic acid.

In summary, SPME holds promise for capturing and concentrating a wide range of malodorous compounds associated with wastewater, with subsequent chromatographic analysis to determine concentrations at levels of parts per trillion by volume to parts per billion by volume in air. It can be used for direct sampling in air or in wastewater, or for headspace sampling above a wastewater sample, and has several advantages over more conventional sampling methods. Headspace sampling appears to be the least problematic of these approaches at this time. Additional research is needed, particularly related to the application of SPME for capturing odor-causing compounds associated with untreated wastewater.

4.5.4 Electronic Nose

Electronic nose systems use an array of nonspecific sensors to provide a "fingerprint," or unique profile, of an odor without determining its actual chemical composition. An odor's fingerprint is established by using sensors with varying affinities for odorous gases. Sensor materials can range from conducting polymers to metal oxides and from surface acoustic wave devices to bioenzyme sensors. Conducting-polymer sensors, which are used to measure changes in electrical resistance across polymers, offer the largest range of selectivity and sensitivity to different compounds (Stuetz et al., 1998). The sensor responses can then be further analyzed using pattern recognition techniques (such as multiple-discriminate and principal-component analysis) or neural network algorithms. The choice of analysis technique is dependent on the amount and nature of information available and the type of information required from the analysis (that is, quantitative or qualitative).

The principal components of an electronic nose system are its sample delivery, the sensor array module, and the data analysis module (Figure 4-8). The delivery system is designed to transfer sampled headspace gas from above a sample material to the sensor array chamber in a reproducible manner. This can be achieved by either generating a headspace above a sample and then introducing the sensor arrays to the headspace (static sampling) or pumping the sample headspace from a sample vessel to the sensor array (dynamic sampling). The different performance characteristics of these sample methods have lead to the development of different electronic nose instruments for different fields of application. A number of electronic nose systems are available from commercial manufacturers throughout the world. These systems can be classified into three application groups: complex laboratory-based instruments, online systems, and portable devices for field measurements.

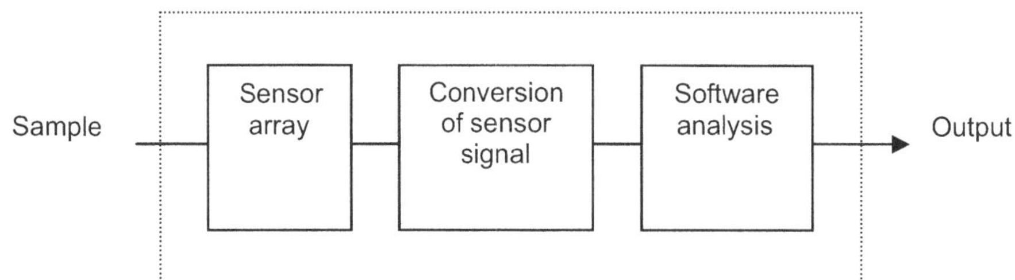

Figure 4-8. Principal Components of an Electronic Nose System.

Two recent studies have shown the potential for electronic nose systems for applications related to odors from municipal wastewater. Stuetz et al. (2000) collected wastewater samples over a five-month period from three locations (inlet works, settling tank effluent, final effluent) at several different municipal wastewater treatment plants. An electronic nose system (an array of 12 polypyrrole-conducting polymers) was used to develop odor profiles from the headspace above wastewater samples. Odor profiles were compared to BOD_5 and OP measurements using canonical correlation analysis. Relatively strong correlations were observed between odor profiles and the wastewater parameters, suggesting an ability of electronic nose systems to detect changes in organic content, and thus odor potential, of wastewater. A particularly strong correlation was observed between BOD_5 and odor profile for raw (untreated) wastewater (from the inlet works), suggesting the possible use of electronic nose systems for monitoring the spatial and temporal odor potential of sewers.

Stuetz et al. (1998) collected Tedlar bag samples from a range of locations at 10 wastewater treatment plants. The contents of bags were analyzed by a dynamic dilution olfactometer to calculate TONs in OUs per square meter. Bag contents were also analyzed to establish odor profiles using an electronic nose system (an array of 12 polypyrrole-conducting polymers). A poor correlation was observed between odor profile and TON when data across all locations were analyzed as a single set. The correlation improved markedly when only data corresponding to TONs less than 4,000 OU/m^3 were analyzed. Furthermore, exceptionally good linear correlations between odor profile and TON were observed when data were grouped into sets for the same treatment works and processes. The linear correlation between odor profile and TON for TONs less than 4,000 OU/m^3 was very good for bag samples collected at inlet works across three wastewater treatment plants, suggesting the possible application of electronic nose systems to predict the TON associated with odors originating from sewers.

The two studies above were summarized by Fenner and Stuetz (1999). Additional applications have included the monitoring of chemical pollutants in wastewater for early warning of process upsets (Bourgeios et al., 2003) and the detection of odor abatement failures (Stuetz and Bourgeios, 2004). The primary benefit of electronic nose technology is that it may eventually provide a means for estimating the effects of malodorous mixtures on human perceptions of odors, and could provide a means for correlating wastewater parameters with fingerprints of malodorous gas mixtures. The primary drawbacks are that the technology does not provide measurements of actual chemicals that compose odors, may not respond at low concentrations for some chemicals with particularly low odor thresholds, and has not been proven in potential through significant and systematic field studies.

4.6 Major Knowledge and Research Gaps

Microbial transformations play a significant role in the formation and subsequent emission of odorous and corrosive compounds into the sewer atmosphere. Under anaerobic conditions, fermentation processes and sulfate reduction are the dominant mechanisms involved in the production of VFAs and hydrogen sulfide, the latter being a dominant source of both odor and corrosion concerns in sewers. The production of these products is dependent on the wastewater source, types of microorganisms present, and operational conditions. The nature of emitted malodorous compounds (types, amount) affects human detection and response to odors associated with municipal wastewater. Unfortunately, much remains to be learned about the effects of mixtures on odor thresholds and perceptions, and analytical methods have yet to be fully developed for standardizing the effects of malodorous chemical fingerprints.

Major knowledge and research gaps that stem from this review are listed below:

1. Past studies related to the corrosion in sewers have focused entirely on the production and release of hydrogen sulfide and formation of sulfuric acid that leads to corrosion of sewer pipes. Additional research related to whether other biochemical reaction products in sewers lead to corrosion, or enhance the corrosion related to hydrogen sulfide releases, is warranted. More detailed information related to corrosion of sewer pipes is provided in Chapter 9.0.
2. Fermentation processes and VFA formation remain areas of uncertainty and are worthy of additional basic (laboratory) research.
3. The concept of odor potential should be expanded to include not only the current state of wastewater but also the potential to form odorants following chemical and biological transformations. This necessarily requires greater knowledge of parameters that lead to transformations and the transformation pathways themselves, and thus a need for more laboratory research.
4. Additional field studies are needed to link sensory perceptions of odor complaints—that is, annoyance—with the chemical composition of these odorants.
5. Two promising technologies that might facilitate such efforts include solid phase microextraction for the concentration of trace concentrations of odorous compounds and electronic nose systems. However, each of these technologies requires further laboratory research, and ultimately field evaluations, to optimize using the technology for odors associated with raw (untreated) wastewater.

CHAPTER 5.0

WASTEWATER COLLECTION SYSTEM MODELING

5.1 Overview

Modeling wastewater collection system odors and corrosion requires adequately representing multiple biological, chemical, and physical processes, such as the following:

- Liquid phase sulfide generation
- Liquid-to-vapor H_2S mass transfer
- Natural liquid phase chemical and biological oxidation of sulfides
- Dissolved oxygen (DO) impacts on sulfide generation
- Liquid and vapor phase bulk transport of sulfides and H_2S, respectively
- Vapor phase H_2S-induced concrete corrosion

Many of these processes have been studied and reported on over the last several decades. The purpose of this chapter is to provide an overview of the information found in the published literature related to recent advances in modeling wastewater collection system odors and corrosion (rather than to provide a detailed explanation of model development or use).

Published information related to various processes is discussed with the intent to increase the reader's general knowledge of factors, parameters, or issues affecting the understanding of specific mechanisms and how best to quantitatively represent them.

Information obtained and reviewed as part of this literature search was not sufficient to allow for a detailed examination of each individual model described or a detailed comparative assessment.

5.2 Models Available

Of all the papers reviewed for this report, 50 related to modeling were identified. Within this subset, six models were cited. The information presented regarding each model varied from a qualitative explanation of the model to a detailed presentation of the algorithms used. A brief description of each model is provided below. Table 5-1 summarizes each model's capabilities as reported in the literature.

5.2.1 Wastewater Aerobic–Anaerobic Transformations in Sewers

This model (or its components) was cited the most often, and thus there is significant information about it available in the public domain. This model was developed in the 1990s as a sewer microbial biochemical oxygen demand (BOD) transformation model (Nielsen et al., 1992; Hvitved-Jacobsen et al., 1998a, b; Tanaka and Hvitved-Jacobsen, 1998).

Yongsiri et al. (2003a, b) presented an overview of the incorporation of a H_2S liquid-vapor mass transfer component into the wastewater aerobic-anaerobic transformations in sewers (WATs) model. Subsequently, Yongsiri et al. (2004a) presented a two-phase model for predicting gravity flow sewer sulfide generation and release into the overlying sewer headspace. In developing this model, the authors considered only sulfide formation, transfer to the vapor phase, and bulk transport within the liquid and vapor phases; potential removal mechanisms such as biological or chemical oxidation were not included. Consequently, the model is reported to be conservative in predicting liquid phase sulfide and vapor phase H_2S concentrations.

Nielsen et al. (2005a) built upon this work and described the addition of the sulfur cycle to the WATS model. Sulfur-cycle components added to the model included sulfide generation, precipitation, biological and chemical oxidation, reaeration, transfer to the vapor phase, and corrosion. For each component discussed, algorithms are provided.

5.2.2 Septicity Prediction and Control Algorithms (SPACA)

Boon et al. (1998) provided an overview of an algorithm developed to predict sulfide generation within force mains (rising mains). Their paper presents results from the model and compares predicted sulfide concentrations to field-measured ones but does not provide the actual algorithms used. Instead, Boon et al. referred to previously published papers that contain the algorithms used. Parameters required in modeling sulfide generation include flow rate, temperature, chemical oxygen demand (COD), pH, and initial DO concentrations.

5.2.3 East Central Interceptor Sewer (ECIS)

Michiels and Salgaonkar (1994) presented the development of the ECIS model, a new model for predicting liquid phase sulfide and vapor phase H_2S concentrations. The authors begin by reviewing two existing sulfide prediction models, SULF.BAS and HS Program, and the limitations associated with each. They report that SULF.BAS is a gravity sewer model that predicts sulfide generation and subsequent liquid phase sulfide concentrations but does not account for liquid-to-vapor mass transfer and does not report headspace H_2S concentrations. HS Program similarly calculates sulfide generation within gravity sewers using the Parkhurst and Pomeroy equation but also calculates concrete corrosion rates associated with the predicted liquid phase sulfide concentration. The limitation of both of these models, according to Michiels and Salgaonkar, is that neither calculates nor provides sewer headspace H_2S concentrations. The ECIS model calculates liquid phase sulfide generation, flux to the headspace, and the resultant H_2S vapor space concentration. Model algorithms and parameter values are provided.

5.2.4 AEROSEPT

Matos and Aires (1995) described the use of a gravity sewer and force main model, AEROSEPT, to predict sulfide concentrations within a collection system in Portugal. Model algorithms were not presented, but information about their sources was provided—for example, use of Pomeroy's (1959) algorithm for pressure main sulfide generation and Pomeroy and Parkhurst's (1977) for gravity sewer sulfide formation. Although Matos and Aires did not compare model-predicted values to field-measured ones, they did present model results for a new interceptor system as a means to highlight the value of the model during system design.

5.2.5 U.S. EPA Sulfide and Corrosion Models

Kienow and Kienow (2004) reviewed the U.S. EPA sulfide model developed by Pomeroy and Parkhurst (1977) and presented in an ASCE (1982) manual of practice. They also referred to and reviewed the U.S. EPA (1974) corrosion model. To enhance these algorithms, Kienow and Kienow described using a turbulence corrosion factor and a crown corrosion factor to account for system-specific corrosion rates induced by turbulence and system design. These two factors' values are provided for various structures and conditions.

5.2.6 INTERCEPTOR Model

Witherspoon et al. (2004) provided an overview of the INTERCEPTOR Model and listed the mechanisms incorporated into it, including sulfide generation, liquid phase and vapor phase bulk transport, oxidation, liquid-vapor mass transfer, and concrete corrosion. The only algorithm presented is related to concrete corrosion (Pomeroy, 1974).

Table 5-1 summarizes which parameters the models' algorithms attempt to address.

Table 5-1. Summary of Models Reviewed.

	WATS	INTERCEPTOR	SPACA	ECIS	AEROSEPT	U.S. EPA Sulfide and Corrosion
Sulfide generation	Yes	Yes	Yes	Yes	Yes	Yes
Sulfide oxidation/removal	Yes	Yes	Yes	Yes	NR	No
Mass transfer	Yes	Yes	No	Yes	Yes	Yes
pH impacts on liquid phase speciation	Yes	Yes	Yes	Yes	Yes	Yes
DO impacts	Yes	Yes	Initial conc. only	No	NR	Yes
Vapor flow rates	Yes	Yes	NR	Yes	NR	No
Corrosion	Yes	Yes	NR	No	Yes	Yes

NR, not reported.

It should be noted that all six of the models discussed are steady state. As such none of them can be used to track the generation and fate of sulfide over the course of transient events such as diurnal flow variation. However, results due to varying conditions can be obtained by conducting multiple model runs with wastewater input parameters (e.g. temperature, or wastewater flow rate) varied.

5.3 Compounds

While it is generally understood that there are potentially several odorous compounds present in wastewater, the primary compound of interest in all modeling papers reviewed was H_2S. Boon et al. (1998), during their discussion of the SPACA model, indicated that other compounds may also be responsible for odors, for example, volatile fatty acids (VFAs) such as acetic, propionic, and butyric acid, and basic compounds such as ammonia, amines, skatole, and indole. However, the authors state that these compounds normally account for less than 5 percent of odors and that H_2S is considered the predominant compound.

Likewise, H_2S was the dominant compound discussed in terms of concrete corrosion and is the compound all published algorithms are developed around in terms of predicting corrosion rates. In addition to discussing the mechanisms of H_2S-induced concrete corrosion, Boehm et al. (1998) reported that other compounds such as nitrates and nitrides may play a role in concrete corrosion. However, the role or mechanism of these compounds is not discussed. It is likely that these compounds play a minor role in concrete corrosion in municipal wastewater collection systems.

5.4 Sulfide Generation

All models reviewed included a term for representing the generation of sulfides within the sewer biofilm. There are two different forms of algorithm used: one for predicting sulfide within gravity flow sewers, where there is a vapor headspace, and another for force or rising mains that have no headspace.

In general terms, sulfide is formed as a result of the anaerobic transformation of biodegradable compounds and the reduction of sulfate. However, as presented by Nielsen et al. (2005b), sulfate is not usually included in sulfide generation equations because it is not rate limiting when it is present in concentrations greater than 5 to 15 mg/L.

Nielsen et al. (1992) provided an overview of four algorithms commonly used to represent force main sulfide generation. Although slightly different in the empirical constants used, the algorithms all utilize temperature and organic material concentration. Some of the equations presented additionally include flow velocity or slope. The form of organic matter concentration used in the various models ranged from soluble COD to BOD. Kienow and Kienow (2004) stated that factors affecting sulfide generation include high temperature, high BOD, flat sewers, and long detention times in fully flowing pipes—for example, force mains, siphons, and surcharged gravity sewers.

Algorithms for gravity flow sewers use similar parameters in describing sulfide generation. However, one important difference is that in gravity flow sewers, there is the potential for reaeration of the wastewater, resulting in the potential to maintain sufficient DO within the wastewater to prevent anaerobic conditions. Therefore, in addition to tracking sulfide, gravity flow sewers must also perform a DO balance. The bulk phase DO concentration required to prevent sulfide generation or accumulation in the liquid phase has been reported to be in the range of 0.2 to 0.5 mg/L (Nielsen et al., 1992; U.S. EPA, 1974).

Several published papers provide information about the best parameter to use in characterizing the nutrients available to sulfate-reducing bacteria. For example, Hvitved-Jacobsen et al. (1999) described a model that has been developed to account for aerobic and anaerobic microbial transformations. Instead of just describing sulfide generation as a function of BOD or COD, they suggest that the anaerobic transformation of sulfate to sulfide is accomplished through mainly the use of fermentable substrates and their byproducts, for example, VFAs. Nielsen et al. (1998) discussed the use of acetate or VFAs as a potential indicator of sulfide generation within force mains because this compound is associated with multiple anaerobic metabolic pathways. A sulfide production algorithm is presented and adjusted for site-specific data by use of an empirical constant used to adjust for different wastewater types—for example, industrial, domestic—and a different temperature adjustment factor. However, the algorithm uses soluble COD as the substrate indicator.

Michiels and Salgaonkar (1994) reported on the use of the sulfide generation rate constant (M') developed by Pomeroy (1959) as an initial starting point for estimating sulfide generation in the ECIS model. The authors report that field measurements used during model calibration resulted in the determination that the actual system-specific M' was six times greater than the default value developed by Pomeroy. These results suggest that it is important to obtain system-specific data so that an appropriate value of M' can be used.

5.5 Dissolved Oxygen

DO is an important component of the sulfide generation phenomenon because at sufficient concentrations it is capable of maintaining an aerobic environment in which sulfide generation will not occur. However, DO is continually consumed by sewer biomass and must be replaced through a gas-to-liquid mass transfer (reaeration). Much of the research on DO has focused on improving the knowledge base associated with these competing mechanisms.

Nielsen et al. (1992) discussed the importance of DO uptake and used the algorithms developed by Parkhurst and Pomeroy (1997) to represent reaeration in sewers. They reported literature values of oxygen uptake ranging from 2 to 20 mg O_2/L/hr in gravity sewers and approximately 0.7 g/hr per m^2 of biofilm surface area within rising mains.

Tanaka et al. (2000) measured several parameters in an operating pressure sewer in Japan and evaluated the use of air injection as a means to provide sufficient DO to prevent septic conditions. They found that sulfide generation was prevented at DO concentrations greater than 1 mg/L.

Gudjonsson et al. (2002) measured DO concentrations within an operating 3.6-km interceptor over a two-month period. They then compared the results of this monitoring to modeling results. The authors accounted for biological utilization of DO in both the bulk phase and within the biofilm in addition to the mass transfer component. The comparison of predicted and measured DO concentrations was reported to have yielded good long-term but not short-term averaging results owing to potential fluctuations in wastewater composition that were not captured within modeling input parameter values.

5.6 Sulfide Oxidation

One method for removing sulfide from the liquid phase is through oxidation. There are two naturally occurring forms of sulfide oxidation, biological and chemical. Chemical oxidation is primarily mediated through oxygen in the bulk phase whereas biological oxidation is carried out by bacteria within the sewer biofilm.

Nielsen et al. (2005b) compared the relative importance of chemical versus biological oxidation of sulfides. They provide references indicating conflicting information on the relative importance of the two forms of oxidation. For example, the authors present information from ASCE (1989) that states that chemical oxidation is slower than biological oxidation, whereas Wilmot et al. (1988) determined that chemical oxidation makes up 44 to 88 percent of sulfide oxidation. Nielsen et al. presented two algorithms developed to predict chemical and biological sulfide oxidation and state that the chemical oxidation rate is dependent on the form of sulfide present and that H_2S is oxidized slower than hydrosulfide ion (HS-).

Nielsen et al. (2004) reported that chemical oxidation is temperature dependent and doubled with an increase of 12°C.

5.7 Mass Transfer

Several papers on the issue of liquid-to-vapor phase mass transfer have been published. Extensive research has been performed in the past on oxygen reaeration of sewers and streams and much of the current H_2S-related mass transfer research has been done developing refined oxygen: H_2S mass transfer correlations.

Nielsen et al. (2005a, b) built upon the oxygen reaeration equation for sewers developed by Parkhurst and Pomeroy (1972) and report that the overall mass transfer coefficient ratio of $H_2S:O_2$ is 0.86.

Yongsiri et al. (2004) reported that the H_2S mass transfer rate (KLA H_2S) is dependent on pH. They provide an algorithm that predicts KLA H_2S as a function of pH (range of 4.5-8.0) and oxygen mass transfer (KLA O_2). For a given pH, the ratio of $H_2S:O_2$ mass transfer was 0.86, the same as that reported by Nielsen et al. (2005a, b). Huisman et al. (2004) described a set of experiments and a new algorithm for calculating sewer reaeration Provided also are a summary of available reaeration equations and methods—direct and indirect—routinely used to measure oxygen transfer. In developing the proposed algorithm, the authors used an inert tracer gas, sulfur hexafluoride (SF_6). The proposed equation indicates that oxygen mass transfer is related to oxygen solubility, the presence of surfactants and three dimensionless parameters (Froude, Reynolds, and Schmidt numbers). The surfactant adjustment factor, which is the ratio of the wastewater mass transfer rate to the clean water mass transfer rate, was reported to be as low as 0.4 but typically in the range of 0.5-0.9.

Yongsiri et al. (2004b) described research done to determine the impact of temperature on mass transfer. The authors present an expression that follows the form of the Arrhenius equation and discuss the importance of separating the effects of pH and temperature on mass transfer when considering the mass transfer rate on the basis of dissolved sulfides, which includes H_2S (the component of sulfides that can volatilize) versus the ionized component HS-, which cannot volatilize. Their research also indicated that conductivity has little impact on mass transfer.

Lahav et al. (2004) presented a new method for predicting H_2S emissions from gravity flow sewers. They develop a model using a velocity gradient as the parameter to describe hydraulic conditions within the sewer. This method was compared to the U.S. EPA (1985) equation using sulfide concentration data collected from an operating sewer in South Africa. Results indicate that their new method performs better than the U.S. EPA method. However, these results should be considered preliminary because other mechanisms such as generation or removal are not included.

Chern and Yang (2004) reported on experimental work developed to better understand oxygen mass transfer across drop structures. The authors describe two zones for mass transfer: the droplet (the zone in which the water is falling and disintegrates into droplets) and the surface (the tailwater surface). The authors varied temperature, drop height, receiving water depth, and flow rate and found that mass transfer rate for both zones increased with drop height and flow rate. Water depth also impacted the surface zone mass transfer but did not impact droplet zone.

5.8 Ventilation

For gravity flow sewers, the flow of vapors within the sewer headspace is an important parameter in completing mass balances for the liquid and vapor phases. However, little research has been done on this topic. Previous works include those by Pescod and Price (1981), who published information related to liquid-drag-induced natural ventilation; Quigley and Corsi (1995), who measured natural ventilation rates in a large municipal sewer; and Olsen et al.

(1997a, b), who presented a mechanistic approach to calculating sewer ventilation on the basis of thermodynamic fundamentals.

When possible, it is best to use actual field data to calculate sewer natural ventilation rates. For example, Michiels and Salgaonkar (1994) reported that the ECIS odor model used air and liquid velocity data collected in an operating sewer to develop a method for estimating natural ventilation rates. Field data collected indicated that the air velocity was equal to 0.05 times liquid velocity. This could then by multiplied by the headspace cross-sectional area to obtain a ventilation rate.

When field data are not available, it is necessary to estimate the ventilation rate. This can be done by assuming either a constant air:water flow rate or velocity ratio. Yongsiri et al. (2004a, b) used a constant air:water velocity ratio of 0.2, citing previous works that indicated the range may vary from 0.2 to 0.3.

Edwini-Bonsu and Steffler (2004) presented a computational fluid dynamics-based model developed to calculate air flow in sewers. The model is compared to previously published models and measured ventilation-rate data. Results of this comparison suggest that the previous models may overpredict the air flow rate in sewers. While the data used for the comparison are relatively sparse, the potential for this model to provide an improved method for predicting natural ventilation rates in sewers is significant.

5.9 Corrosion

Two notable papers related to H_2S-induced concrete corrosion were published. Kienow and Kienow (2004) reviewed an ASCE (1989) algorithm that was developed to predict the quantity of concrete thickness necessary to prevent pipe failure and that involves using the crown corrosion and the turbulence corrosion factors. They reviewed factors impacting concrete corrosion rates and pipe life, including the level of turbulence as accounted for by the use of turbulence correction factors, concrete alkalinity, and concrete cover thickness z, which can be specified by the designer. The primary factor impacting rate of corrosion was stated to be turbulence which impacts the amount of H_2S stripped from the wastewater. According to the authors, concrete with limestone aggregates has four to five times the alkalinity and resistance to corrosion as granitic aggregate concrete. The authors also discuss the mix factors that may impact corrosion rates including cement types, fly ash addition, and high alumina cement.

Boehm et al. (1998) developed a corrosion model and evaluated relative importance of several variables including: concrete surface pH, concrete alkalinity, dissolution rate of calcium carbonate to calcium sulfate. While the focus of the paper is on H_2S's role in concrete corrosion, Boehm et al. also pointed out that nitrates and nitrides may cause corrosion. However, the relative importance or the manner in which these compounds cause concrete corrosion is not discussed. Boehm et al. setup a hypothetical example using 5 ppmv H_2S as the headspace concentration and varied concrete pipe surface pH from 1 to 4, the dissolution rate from 0.0084 to 0.84 cm/day and concrete alkalinity from 0.2 to 10 mol/L. Concrete loss was predicted to range from 0.3 to 27 cm over a 25-year period.

5.10 Relationship Between Sewer Headspace H_2S Concentration and Corrosion

Several models exist that allow for the prediction of liquid phase sulfide generation and transport within sewer systems. Additionally, some of these models attempt to facilitate the prediction of sewer corrosion through implementation of empirical formulae that are based primarily on liquid phase sulfide concentration data.

Given that corrosion is related to sewer headspace H_2S concentration, it is recommended that an algorithm(s) be developed to account for this relationship. This would be an improvement over the existing methodology as this new method of prediction of corrosion could be completed using the following steps:

1. Liquid phase sulfide generation
2. Liquid-to-vapor mass transfer of H_2S
3. Vapor phase H_2S fate modeling

Steps 1 and 2 already exist and account for the generation and fate of sulfide within the liquid phase, including bulk transport downstream, chemical or biological oxidation, and interphase mass transfer.

Step 3, however, requires additional focus. It is recommended that research be completed to allow for the development of algorithms that correlate corrosion to several different factors, such as the sewer unwetted wall area, material of construction, bulk flow of air, and headspace H_2S concentrations.

A research plan describing this proposed work in more detail is included in Chapter 10.0.

5.11 Summary and Conclusions

Several wastewater collection system models have been developed to facilitate the assessment of sulfide generation and potential release into the sewer headspace atmosphere. While several of the models rely on similar methods for approximating sulfide generation and liquid-to-vapor mass transfer, the methods for predicting ventilation rates and inclusion of concrete corrosion as a H_2S loss term may not be consistent. Improved understanding of the fundamentals associated with these two areas—natural ventilation of sewers and quantification of vapor phase H_2S loss associated with concrete corrosion—are therefore considered key areas for future research.

CHAPTER 6.0

SEWER VENTILATION

6.1 Literature Search Background Summary

Nineteen papers were reviewed for pertinent information on sewer ventilation. The primary screening criterion for determining a paper's relevancy was a significant discussion of or focus on sewer air movement, headspace pressurization, or sewer emissions. Thus, some of the papers discussing treatment of sewer off-gas, chemical treatments in collection systems, or dispersion modeling were screened but not reviewed in detail. Twelve of the 19 papers were significant in that they had data related to characterizing or handling sewer headspace air movement and pressure conditions. Generally, each of the 12 papers fell into one of three categories:

- Case studies related to resolving agencies' collection system odor problems (nine papers)
- Research studies, focusing to various degrees on pilot-scale and full-scale ventilation using tracer gas methods (two papers)
- Doctoral thesis describing sewer ventilation using a computation fluid dynamics model (one paper)

Seven papers briefly touched on sewer ventilation but focused on odor treatment aspects rather than the dynamics of airflow and pressure development within a sewer headspace.

All papers were dated between 1994 and 2005.

6.2 Introduction and Background

Wastewater flow in gravity sewers produces a cocurrent flow of air in the headspace above the liquid. In many interceptors, air movement and sewer-headspace pressurization are negligible and warrant minimal analysis or design considerations to avoid interceptor odor emissions. In cases when sewer odor emissions are potentially significant, chemicals can be added to control liquid phase odorants that may otherwise be released to the sewer atmosphere. However, chemical controls are not always totally effective, especially in critical receptor areas, leaving headspace ventilation and treatment as the only practical means of totally controlling sewer pressure and interceptor odor problems. Based on the literature survey of case studies, interceptor systems that are prone to fugitive emission problems are typically long (from 3 miles up to 50 miles), large-diameter (42-132 inches) trunk sewers. These systems often exhibit headspace air pressures between -0.2 and +0.2 inches of water column (w.c.) In certain situations, pressure that has developed behind full pipes and siphons can exceed +1.0 inch of w.c. In air movement scenarios, these pressures are relatively low; they can, however, produce foul-air outleakage of several hundred cubic feet per minute (cfm) in poorly sealed manhole covers. Serious fugitive emissions problems have been associated with junction structures, siphons, and highly variable liquid flow conditions, which can greatly affect the movement of air through the interceptor system as well as produce greater hydrogen sulfide stripping from the wastewater, due to liquid turbulence.

The nature and extent of air exchange (ventilation) between the ambient atmosphere and sewer headspace has never been adequately characterized. It is clear that the potential for fugitive odors is a direct consequence of headspace air pressurization, which is controlled primarily by airflow resistance and physical interceptor characteristics. Significant pressurization is often associated with constricted headspace, which restricts airflow through the sewer pipe, causing emissions from manhole covers and in upstream lateral sewer connections. Full-pipe or unjumpered siphons are special examples of such restricted airflow conditions. Rapid wastewater flow tends to produce greater air movement due to greater friction drag from the liquid on the headspace air. This results in potentially higher headspace air pressure. However, greater air movement also tends to increase headspace air turnover, thus reducing odorant concentrations and potentially lessening corrosive tendencies in the pipeline.

Since most collection systems are closed systems designed to carry only liquid flow, the engineering effort given to quantifying headspace air movement in designed interceptor systems has been relatively low. In the theoretical analysis and case studies reviewed, there was no case in which a definitive analysis of sewer air movement had been conducted as part of a new interceptor system design. However, there were several case studies in which evaluations directly related to existing interceptors with pressurization or fugitive emissions problems were conducted. Most of the theoretically based papers and analyses focused mostly on developing and quantifying parameters affecting airflow relative to liquid flow. The case studies tended to focus on measured airflow or pressure conditions and either analyzed the efficacy of foul air extraction or developed a basis for determining the amount of sewer foul air extraction needed to resolve odor conditions in site-specific circumstances.

6.2.1 Theoretical Dynamics of Air Movement in Wastewater Conduits

The primary force driving natural air movement in water conduits is friction between the sewer headspace air and flowing wastewater, whereas the natural resistance to flow is friction between the walls and airflow. As one would expect, this results in a theoretical air velocity profile reflecting maximum airflow near the water surface and minimum airflow near the wall surface. One of the earliest and more detailed references for sewer ventilation is work done by Pescod and Price (1981). Their work is referenced in many of the papers that were reviewed. They identify the following five factors influencing sewer ventilation:

- Wind eduction
- Wastewater drag
- Rise and fall of wastewater level
- Temperature differentials
- Barometric pressure differentials

They determined that the wastewater level, temperature, and barometric effects can be dealt with analytically. Wind effects and wastewater drag are not amenable to analytical treatment, and therefore these factors are the focus of their research and testing. A unique aspect of their work is their focusing the majority of their efforts on wind eduction, using carefully crafted stack heads to make maximum use of ambient wind to induce sewer airflow through the stack and headspace. This approach to sewer ventilation is not widely practiced in the United States because of the attendant odor problems that would likely develop from untreated sewer emissions.

The Pescod and Price work included both laboratory-scale tests and full-scale system evaluation. The laboratory tests evaluated wind eduction and liquid drag separately. The wind eduction tests were done with a wind tunnel, and the liquid drag tests were done by operating various liquid velocities and depths in a 300-mm pipe, with airflow measured by thermal anemometers. Their full-scale system evaluations analyzed the combined effects of wind eduction and wastewater drag.

Their wastewater drag findings are shown in Figure 6-1. This represents idealized airflow velocity profiles for one test in which average liquid velocity is 0.8 m/s.

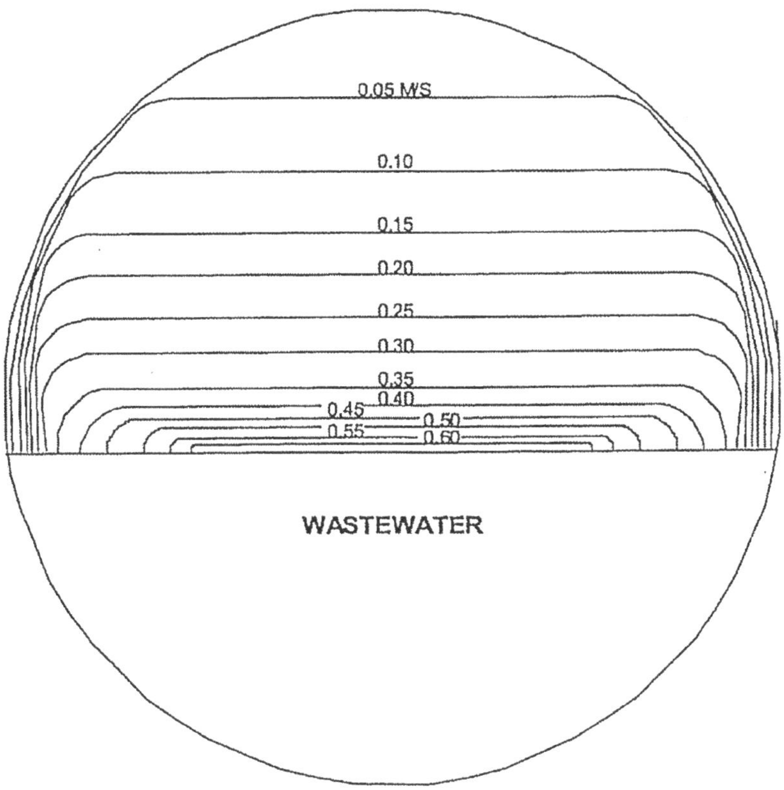

Figure 6-1. Idealized Velocity Contours.

Overall, their data covering nine separate drag-induced flow tests showed considerable scatter, making correlations difficult. However, they did relate average air velocity to water surface velocity, and to both air space perimeter and air space "hydraulic" radius. The data reflected an apparent maximum air velocity limit, regardless of liquid velocity and depth considerations. Although the applicability of their research is limited by the small pipe size and modest liquid velocity (less than 0.6 m/s), their work is referenced in many sewer ventilation papers. An often-cited rule of thumb from this work suggests that airflow velocity will be 35 to 50% of liquid velocity in a half-full sewer. It is clear from their analysis (which included tests at various flow depths) that airflow above the wastewater can vary significantly depending on the width of the liquid surface, since the liquid surface is the prime liquid drag motive force to move the air along with the wastewater. Their work suggests that the point of maximum airflow is a condition of a half-full sewer, which maximizes the surface exposure to the above headspace.

The primary limitation of the Pescod and Price (1981) work is that it develops data on the basis of a relatively small pipe with a liquid flow at a relatively modest velocity; most collections systems are far more complex. Hydraulic jumps, vertical drops, large junction and diversion structures, inverted siphons, and pump stations add significant complexity to the factors that govern sewer ventilation. If wastewater flow is fast and turbulent or if there are large hydraulic jumps or vertical drops, liquid surface area and velocity increase significantly, greatly increasing drag-induced flow in the surrounding air. Conversely, slow-moving quiescent streams will exert minimal drag on the air and move relatively small volumes of air.

6.3 Sewer Ventilation and Pressurization Case Studies

Most published work on sewer ventilation is related to case studies looking at emissions control in specific systems. These papers analyze deep tunnel systems, large-diameter trunk interceptors, and special cases involving industrial sewer volatile organic compound (VOC) emissions. Some studies use laboratory evaluations of sewer ventilation and compare results to full-scale systems.

6.3.1 Deep Tunnel Systems

Deep tunnel systems are a special case in sewer ventilation. These systems employ vertical drop structures in which the wastewater falling vertically into the tunnel induces large amounts of airflow. Typically, various energy dissipation structures and air-relief vent pipes are provided with these systems. Three papers in the literature search focused on deep tunnel systems that were evaluated from a ventilation perspective, focusing on mitigating specific fugitive odor emissions.

A deep tunnel system in Columbus, Ohio, was studied by Sorenson et al. (2000). The system that was evaluated consists of six drop structures to a tunnel 72 inches to 84 inches in diameter. Key findings included fairly consistent diurnal pressure variations at a drop structure. Pressures ranged from +0.2 inches to -0.15 inches of water column (w.c.). To provide fugitive emissions control, airflow requirements were estimated from both pressure data and wastewater flow data. Airflow estimates developed from pressure data ranged from a minimum of 1,300 cfm to 10,000 cfm at various points in the system. Airflow developed by correlation to wastewater flow (pipe half full and air velocity equal to half the liquid velocity) produced flow estimates of 2,500 to 5,500 cfm applied to three locations along the system.

A deep tunnel system in the Northeast Ohio Regional Sanitation District was evaluated by Smith and Klunzinger (2001) to assess pressures with and without odor control system fans operating. The diameters of the pipes in this system range from 42 to132 inches, and five biofilters are installed to treat foul air exhausted from the interceptors. Their paper evaluated the level of "baseline" pressure that exists in this system without any forced ventilation from the biofilter fans. These data were compared with pressures using various combinations of biofilter fans. Ten manholes were tested; data reflect that seven of the 10 locations exhibited negative headspace pressure, even with no fans operating. Pressures ranged from -0.78 inch of w.c. to +0.12. Six of the 10 locations averaged between -0.01 and -0.10 inch of w.c. With the fans on, negative pressures increased; however, one manhole developed positive pressure, and another manhole that had been under positive pressure remained so after the fans had turned on. The study evaluated these data in the context of the system design and locations where restrictions to air movement were likely. At one key pressurization point, an estimated 6,500 cfm of air was flowing from the convergence of three sewers and accounting for pressurization at that point.

Odor problems from a deep tunnel system in Austin, Texas, were evaluated by Pope et al. (1998). The essence of their work is that the tunnel ventilation problems and fugitive odor emissions were influenced almost entirely by a downstream pumping station's wet well odor control system. The findings were that influent pumping required liquid levels that submerged or nearly submerged the inlet pipe. This process sealed the air space in the influent pipe, thereby limiting airflow to the pump station odor control system and stopping the natural draft and ventilation through the tunnel system.

6.3.2 Large-Trunk Sewer Evaluations

Ventilation and headspace pressure conditions were studied in papers covering three large-trunk interceptor systems. These systems tend to have large-diameter pipes with large available headspace for airflow.

The Potomac Interceptor, which links jurisdictions in northern Virginia with a wastewater treatment plant in Washington, D.C., was evaluated to determine feasible approaches to odor control, including both chemical treatment and foul air extraction (Trypus et al., 2001). This interceptor is a 50-mile-long regional gravity interceptor that is vented along most of its routing to promote the exhausting of sewer gases and their replacement with fresh air. Trypus et al. evaluated forced air exhaust from the headspace using a 5,000-cfm blower. On the basis of pressure measurements from the field pilot study, it was determined that at a 7,300-cfm air exhaust rate would produce pressure influence approximately three miles upstream and 1,000 feet downstream of the exhaust point. The 7,300-cfm exhaust rate was developed by applying a 0.7 factor to the liquid velocity to determine air velocity. The 0.7 factor is an average of data from Thistlewaite (1972) and Joyce (2000). In the 84-inch pipe being evaluated, this resulted in a 6,000-cfm airflow, which was then adjusted upward based on pressure observations from the blower air extraction test.

A Phoenix, Arizona, study on the Southern Avenue Interceptor trunk sewer (diameters of 42 to 72 inches) and Salt River Outfall sewer (90-inch section) examined forced air extraction to evaluate depressurizing the sewer headspace (Davidson et al., 2002). Figure 6-2 shows a portable fan skid and temporary flexible duct used to conduct this evaluation hooked up to a manhole. A variety of tests was conducted, with flow rates varying between 2,700 and 5,400 cfm. Table 6-1 summarizes pressure data, showing the sewer headspace shift from positive to negative pressure and pressure influence of several miles. The work in Phoenix evaluated headspace H_2S concentrations but was unable to correlate these values to sewer ventilation rate. The work also confirmed that headspace pressure follows a diurnal pattern, a fact mentioned elsewhere in the literature.

Figure 6-2. Air Extraction Skid for Sewer Ventilation Testing (Phoenix, Arizona).

Table 6-1. Phoenix, Arizona, Area Sewer Pressure Variation with Forced Ventilation.

Location	Sewer Diameter (inches)	Airflow (cfm)	Normal Pressure[a]	Pressure with Air Extraction	Distance of Pressure Influence on Sewer (miles)
Tempe, SAI	42	2,700	+0.10	-0.80	4
Phoenix, 51st Avenue Siphon, SAI	72	4,000	+0.50	-0.50	5
Phoenix Airport Parking Facility, SRO	90	5,200	+0.02	-0.13	4.5

[a]Average pressure experienced in sewer headspace in unventilated conditions.
SAI, Southern Avenue Interceptor.
SRO, Salt River Outfall.

A collection system odor and corrosion study in Los Angeles by Hagekhalil et al. (1994) identified a number of factors typically contributing to sewer odor emissions and odor complaints (unvented siphons, wastewater turbulence, etc.). As part of the study, the authors measured sewer headspace velocity with a vane anemometer in a 78-inch pipe. Their data indicated an air-to-liquid velocity ratio of 0.04, with no measurable velocity at 8 inches from the liquid-air interface. Liquid velocity was estimated at 5 fps. These results suggest previous work by Pescod and Price (1981) would estimate air velocity a factor of 10 too high, if applied to a larger diameter sewer.

A study of a large-diameter trunk sewer in Clark County (Las Vegas area) evaluated hydraulic conditions in a steeply sloped sewer with several hydraulic jumps (Pai et al., 2000). The evaluation concluded that hydraulic jumps would produce significant pressurization in upstream sections of the interceptor. This analysis led to a system requiring an external odor control duct running in parallel for 3 miles to the main treatment plant. A 5,000-cfm booster fan system was installed to convey foul air through a 36-inch duct running parallel to the interceptor.

6.3.3 VOC Emissions Evaluations

Two papers included a discussion of sewer ventilation with respect to VOC emissions. In these evaluations, sewer ventilation estimates were made to establish VOC emission quantities, and sewer pressurization was not the major issue.

One VOC emissions study evaluated out-gassing through four perforated manhole covers within the lower part of a 1.6-km sewer reach (Corsi et al., 1995). Pipe diameters ranged from 0.9 to 1.2 m, and pipe slopes varied from 1.0% in the upper part of the reach to 0.25% in the lower part. Thermal anemometers were used to estimate sewer out-gassing velocities. The combined out-gassing flow rate ranged from 1,300 to 2,300 m^3/hr during three different testing events. During a fourth testing event, the furthest downstream manhole exhibited flows ranging from a low of 140 m^3/hr to a high of 590 m^3/hr.

Melcer et al.(1994), in a study evaluating ventilation and its impact on VOC emissions in collection systems, modeled conditions needed to maintain a sewer headspace at less than 10% of the low explosive limit (LEL) at given liquid phase concentrations of VOCs. In other words, this study modeled the highest potential liquid phase VOC concentration that could be tolerated to maintain the sewer atmosphere below 10% of LEL. This study confirmed through modeling (with Toxchem+) that an order-of-magnitude increase in ventilation (air velocity ranging from 0.5 to 5 times the liquid velocity) allowed liquid phase VOC concentrations to be six to eight times higher for the compounds under study (the benzene, ethylbenzene, toluene, and xylene family) without exceeding a 10% LEL condition.

6.3.4 Theoretical Analysis and Pilot-Scale Tracer Evaluations

Three papers offered pilot testing and theoretical perspectives on sewer ventilation rates that were subsequently compared to full-scale sewer segments. Sewer airflow algorithms were analyzed in these studies and applied to full-scale studies.

Monteith et al. (1997) conducted pilot tracer gas ventilation testing and followed this with full-scale tracer testing. Many of the parameters in their study were those known to influence airflow and were originally investigated by Pescod and Price. These included water flow, water temperature, and wind velocity over sewer openings. They attempted to fit their data to an empirical model:

$$v_g = a v_w^{\,b} + c w^d \qquad (6.3.4\text{-}1)$$

where

v_g = headspace gas velocity

v_w = water velocity in the reach

w = surface wind velocity over reach opening

Their optimized solution to the model that was developed from pilot plant tests utilizing tracer gases is

$$v_g = 0.236 v_w^{\,1.12} + 0.0288 w^{0.88} \qquad (6.3.4\text{-}2)$$

Comparing these models with tracer test results on a full-scale system indicated the model had overpredicted gas velocities by a factor of 3 to 5.

The second paper (Parker et al., 2001), a tracer study for the Regional Municipality of Ottawa-Carleton, employed both pilot- and full-scale tests to assess the impact of wastewater drag and wind on sewer ventilation. In the pilot study, average air velocity was found to increase with the product of the water surface width and the liquid velocity divided by the hydraulic radius of the headspace. The full-scale tests reflected highly variable headspace velocity (3.8 to 31.5 m^3/min). Air velocity developed from the field data correlated poorly with the above formulas—ranging from 3 to 30 m/min for one specific value of surface width, liquid velocity, and hydraulic radius.

The third paper was developed primarily from the doctoral thesis of Edwini-Bonsu (2004). This work addressed sewer ventilation from a computational fluid dynamics (CFD) approach and used much of Pescod and Price's original data for evaluation purposes. It partitioned the algorithms representing flow by wastewater drag into both turbulent and laminar flow regimes. The doctoral thesis also included modeling a portion of the City of Edmonton's interceptors to validate flow conditions in that system. The system in question has several drop structures and is ventilated with a foul-air treatment system.

The validity of the CFD model was limited owing to very little available input data for the City of Edmonton system under study. Physical verification of the model predictions was also limited. Additionally, another limitation of the model is that it assumes an average flow condition, which is not representative of a typical sewer system, in which flow and depth varies hour to hour. The potential advantage of the CFD approach is that it appears to use a more sophisticated and potentially more accurate, parameter to estimate liquid-induced friction drag from the liquid interface to the air space. Application of the model has produced estimated airflows that are slightly lower than the simpler correlations that were developed from Pescod and Price.

6.4 Engineering Perspectives of Sewer Ventilation and Odor Control

Although many of the case histories and much of the research provided unique perspectives on sewer ventilation and pressurization, few of the studies developed a logical discussion of obvious cause-and-effect problems relating to this subject and, more importantly, the significant design issues that follow from an analysis of sewer ventilation. The following summarizes issues whose discussion in the research papers reviewed was limited.

A key factor relating to pressure development is the physical structure of the sewer and the available headspace area under various flow conditions. If adequate headspace is available to accommodate the induced airflow, pressurization may be minimal. However, if the headspace is constricted, greater air pressure develops. In cases where the sewer is completely full (surcharged areas, unjumpered siphons, etc.), all available energy imparted to the headspace is reflected as pressure rather than air movement. In these cases, it is highly likely that foul air will migrate up through manhole covers and laterals—and even through side sewers into customers' buildings—to find relief points. In many cases, the odors that are escaping in upstream laterals are addressed by sealing manhole covers. However, since foul air can no longer be relieved at these points, the pressurization tendency actually increases. A secondary problem arises because with less airflow in the headspace, H_2S concentrations increase, so that fugitive emissions contain higher concentrations of H_2S with much greater odor and corrosion potential over a larger area.

The only effective solution to resolve pressure conditions is to mechanically extract air from the headspace and treat it. This approach raises a number of technical design questions concerning how much flow capacity is required and at what point in the system the headspace air should be extracted to achieve maximum pressure reduction benefits. In larger interceptors, the airflow question has been addressed by using the Pescod and Price rule of thumb on air movement (airflow velocity of one-third to one-half the liquid velocity). However, some data suggest that Pescod and Price's airflows overstate actual conditions. A second major parameter relates to how much pressure reduction can be achieved in a given section of interceptor system. This parameter is significantly influenced by the interceptor system construction, including location and size of junction structures, leak-tightness of manhole covers, vertical drop of wastewater from laterals, and the presence of siphons. The Phoenix gas extraction tests described earlier (Davidson et al., 2002) demonstrated headspace pressure reduction up to 5 miles on one interceptor that had relatively well sealed manholes and no major lateral connections. However, gas extraction tests on another section of interceptor demonstrated very little pressure reduction at a major junction structure less than 1 mile from the extraction point. The study for Northeast Ohio Regional Sanitation District tunnel system described above (Smith et al., 2001) found considerable pressure variances that depended on how many biofilter treatment systems were operating. This project also pointed out that the hydraulic design, size, and configuration of junction structures can effectively block the pressure influence of downstream biofilters on upstream interceptor segments.

Mechanical air extraction and treatment raises other issues relative to how and where replacement air is inspirated in the interceptor system. Since a typical goal of mechanical air extraction is to draw a large section of pipe headspace to negative pressure, points near the draw location where air can enter the sewer should be sealed to minimize air "short-circuiting" close to the treatment system. This raises further issues concerning sealing the system or perhaps installing air baffles to partially restrict air movement in various interceptor segments.

6.5 Summary of the Knowledge Base

Several broad conclusions can be drawn from this review of sewer ventilation literature:

- That conceptual factors, such as wastewater drag, influence sewer ventilation is generally agreed upon by most researchers. However, the success in quantifying these factors to yield useful guidelines that agree with field data or full-scale tracer studies has been very limited.
- Pressurization and foul-air fugitive emissions are critical problems in several systems, and several papers reflect case studies of these systems. However, systematic design approaches for dealing with sewer pressurization are notably absent, and many foul-air extraction systems seem to be designed by arbitrary rough estimates and airflow approximations. There is very little empirical methodology employed with these foul-air extraction systems.
- Sewer airflow algorithms are studied in some of the papers. However, the value of these algorithms is questionable because both airflow and pressure will be significantly influenced by the physical characteristics of a particular system. Therefore, interceptor system characteristics play as much of a role in sewer ventilation as the factors that impart airflow in the headspace.

- Data from field studies are very difficult to interpret and often widely scattered. It is likely that continually changing flow conditions, common in any interceptor network, produce greatly differing airflow conditions depending on time of day. Diurnal variances in flow make it difficult to obtain consistent and repeatable data. This problem is aggravated by the rather short duration of most field evaluations that were cited.
- None of the research has produced data showing the relationship between sewer ventilation and odorant concentration. One paper (Davidson et al., 2002) attempted to evaluate this relationship, but the data were inconclusive.
- One paper (Hagekhalil et al., 1994) touched upon the issue of "microdispersion" modeling from manhole odor leaks. Dispersion is not a factor relating to sewer air movement and pressurization, but as a practical matter, the dispersion of odors as they migrate from manhole leaks is an important determinate in whether odors will be objectionable—particularly for systems with lower flows and minimal pressurization tendencies. The dispersion element provides perspective on when an odor emission point becomes a source of complaints.

6.6 Research Gaps

Despite the obvious importance of sewer ventilation to collection system odor and corrosion problems, there has been little progress made toward fundamentally understanding the competing factors that affect ventilation. Factors required to determine sewer ventilation rates and relate them to changes in wastewater flow, headspace above the liquid, or physical structure of sewer systems are needed. Improved knowledge in this area is critical to advancing integrated solutions to sewer odor and corrosion problems. Research has tended to evaluate specific collection systems and focus on solving site-specific odor problems. Full-scale field data often vary considerably from predicted values developed from laboratory pilot tests. None of the previous work attempts to define sewer headspace ventilation in terms similar to airflow in a duct and then relate those parameters to wastewater flow and interceptor characteristics. Furthermore, collection system leak-tightness, while being a critical ventilation parameter, has not been assessed in any of the work thus far.

The objectives of the proposed research are to accomplish the following:

- Develop a database that can be used to evaluate existing and future models related to sewer ventilation.
- Evaluate existing models for the prediction of sewer ventilation.
- Investigate relationships among sewer headspace pressure, airflow, wastewater flow, and geometry of the headspace.
- Modify, as appropriate, the models described in the research literature or develop new models on the basis of the results of further testing and investigations.

Within the context of the various ventilation parameters, project goals should include evaluating the implications for interceptor system fugitive emissions, odor concentrations, corrosion potential, and control method approach.

CHAPTER 7.0

LIQUID PHASE ODOR AND CORROSION CONTROL

This chapter summarizes the literature research findings for controlling hydrogen sulfide release by adding chemicals to the wastewater. Impacts to non-H_2S odors are also considered and summarized where information was found in the literature.

7.1 Literature Search Background Summary

The research team located one hundred papers providing pertinent information on liquid phase treatment in collection systems. Some covered multiple topics, comparing various liquid phase control approaches. The numbers of papers by topic are as follows:

- Aeration using either air or oxygen injection (9)
- Chlorine oxidation (8)
- Potassium permanganate oxidation (1)
- Hydrogen peroxide oxidation (18)
- Iron salts for precipitation (29)
- Nitrate inhibition (34)
- pH adjustment (16)
- Microbial inhibitors, enzymes, and bacterial treatments (17)

Figure 7-1 shows a distribution in terms of the percentage of literature identified.

Virtually all liquid phase treatment topics were covered, but the bulk of the literature focused on either adding nitrates to inhibit sulfide formation, using iron salts to react with already formed sulfide, or adjusting pH for shocking slime layers and shifting dissolved sulfide species to ionic forms that are not released to the headspace.

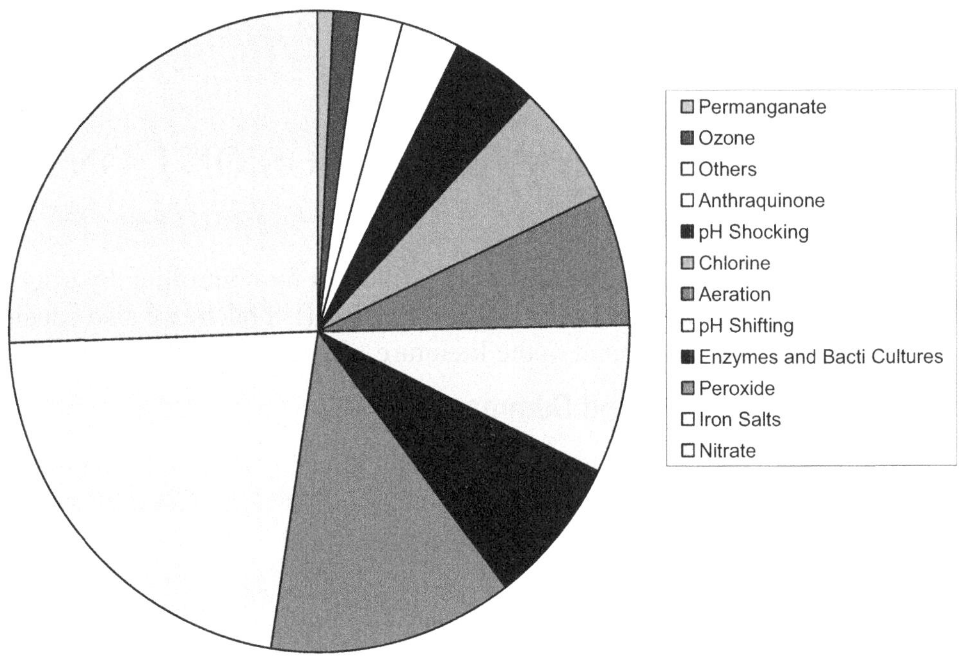

Figure 7-1. Distribution of Papers by Liquid Phase Treatment Topic.

Figure 7-2 shows the distribution of papers among published, peer-reviewed journal articles, conference proceedings, and gray literatures studies.

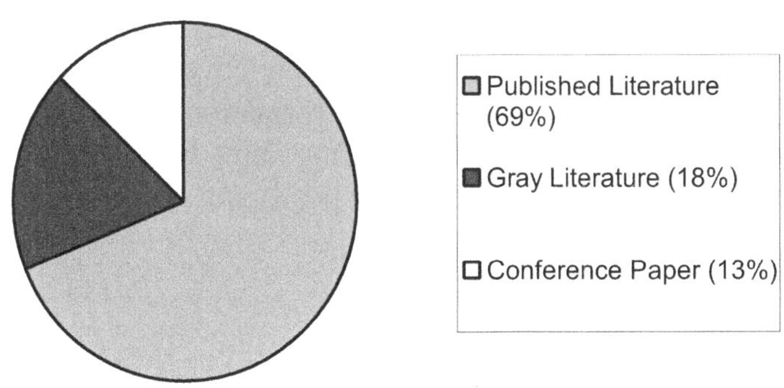

Figure 7-2. Literature Distribution by Source.

The bulk of the identified literature (69%) came from published journal articles, with the remainder split nearly evenly between conference papers and gray literature. Papers and literature covered experiences as far back as 20 years.

Additional valuable information is available from well-established reference books and texts, such as the following:

- WEF (2004) Manual of Practice 25, *Control of Odors and Emissions from Wastewater Treatment Plants*
- WEF (1995) Manual of Practice 22, *Odor Control in Wastewater Treatment Plants*
- U.S. EPA (1985) Design Manual 625, *Odor and Corrosion Control in Sanitary Sewerage Systems and Treatment Plants*
- *Hydrogen Sulphide Control Manual* (Melbourne Water, 1989)

7.2 Introduction and Background

Although various organic and inorganic compounds can cause domestic wastewater odors, hydrogen sulfide was most often identified as the predominant odor- and corrosion-causing compound of concern, and as such, many liquid phase treatment articles focused on removing it.

Most articles focused on the collection system, but some discussed downstream impacts to the receiving wastewater treatment plant or the use of chemicals at the plant's entrance.

Hydrogen sulfide is often released into the gas phase when dissolved sulfide is present. Sulfide is generated within the slime layer attached to sewer pipe walls because of anaerobic biofilm activity. The bacteria responsible primarily for sulfide formation, "sulfate-reducing bacteria," use sulfate (SO_4^{-2}) as their terminal electron acceptor, reducing it to sulfide (H_2S, HS^-, and S^{-2}). The sulfide then diffuses through the biofilm to the bulk liquid phase. Once in the liquid stream, H_2S can volatilize into the gas phase, where it can become an odor and potential corrosion problem. Figure 7-3 is a schematic drawing of this process.

Liquid phase control is a method of odor and corrosion prevention by which chemicals are introduced into the wastewater to react with dissolved sulfides or prevent them from being formed, thereby preventing the release of H_2S into the vapor phase. Various chemicals are potentially suitable for liquid phase odor control. Table 7-1 summarizes generally established methods.

Wastewater characteristics and sewer configuration determine which methods are practical for a given situation. Important parameters include wastewater flow rate, temperature, pH, oxidation reduction potential (ORP), biochemical oxygen demand (BOD), process or sewer detention time, location of odor-releasing points (manholes, pump stations), and constraints imposed by downstream treatment processes. The type of odor-causing compound might also prove to be important.

Cost and practicalities will determine which method is selected. Factors influencing the cost of liquid phase treatment include chemical dose and regional chemical prices. Practicalities include chemical availability and difficulty of transporting, storing, and applying it. The following sections summarize each liquid phase odor control method and highlight the findings from the literature search completed for this study.

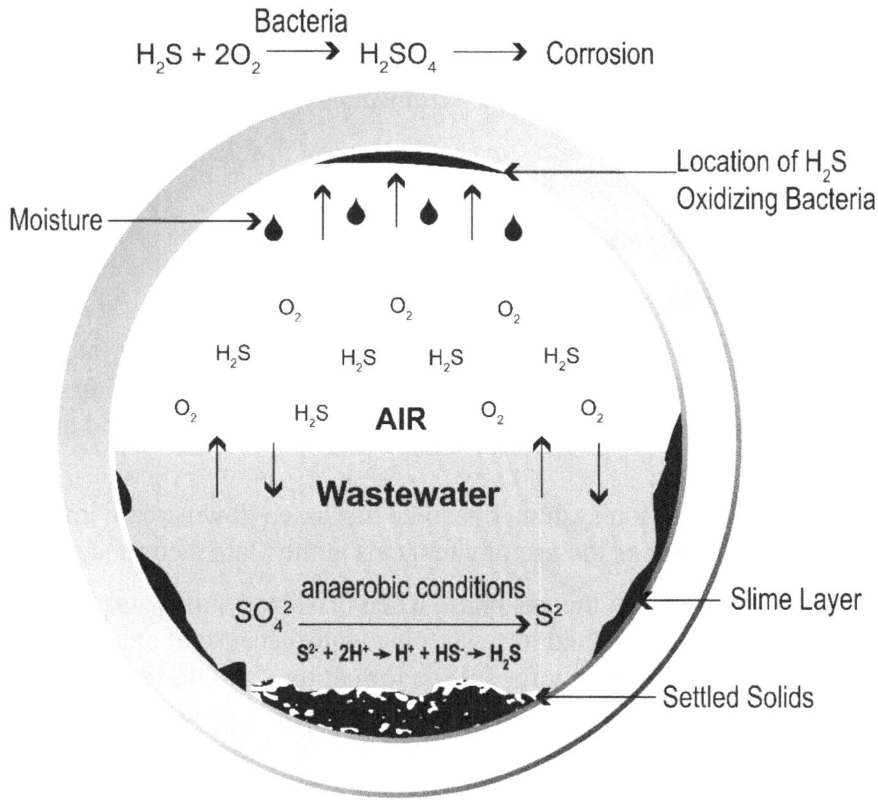

Figure 7-3. Sulfide Transformations in a Gravity Sewer.

Table 7-1. Liquid Phase Odor Control Methods.

Method	Description
Oxygenation and aeration	Ambient air or pure oxygen is injected into the waste stream to oxidize sulfides or to maintain an aerobic biofilm.
Chemical oxidation	Chemical oxidizing agents are added to the waste stream to oxidize dissolved sulfide to sulfate, which remains in solution.
Sulfur precipitation	Metal salts are added to the wastewater stream to form a metal sulfide precipitate with the dissolved sulfide, thus removing it from the waste stream.
Nitrate addition	Nitrate salts are added to the wastewater stream, displacing sulfate as the preferred electron acceptor for anaerobic bacteria and preventing sulfide formation.
pH adjustment	The pH of the waste stream is changed to inhibit growth of H_2S-producing bacteria or to drive volatile H_2S to soluble ionic HS^-.
Biological treatment	Process changes or microbial cultures are used to affect the ability of bacteria to produce H_2S.

7.3 Aeration and Oxygenation

7.3.1 Technology Background

Oxygenation and aeration prevent odors by maintaining a layer of aerobic bacteria on the sewer pipe walls and in the wastewater itself. If a sufficient amount of dissolved oxygen is present in the bulk liquid phase, then the outermost layer of the biofilm can support aerobic bacteria capable of biologically oxidizing sulfide as it diffuses outward through the biofilm. Because of biological oxidation, sulfide would not tend to diffuse into the bulk liquid stream.

However, when dissolved oxygen concentrations within the bulk phase are close to zero, sulfide generated in the biofilm is able to accumulate in the bulk liquid. In addition to preventing sulfide from leaving the anaerobic layer, dissolved oxygen in sufficient amounts can oxidize the sulfide in the bulk wastewater stream. At dissolved oxygen concentrations greater than 1.0 mg/L, oxidation can reportedly keep pace with sulfide production and prevent sulfide buildup (U.S. EPA, 1985).

Methods of adding air or oxygen include the following:

- Direct air injection into force mains
- Use of Venturi aspirators in side streams to lift stations or force mains
- Use of air lift pumps
- U-tube aeration
- Pressure tank injection into gravity lines
- Use of downflow bubble contactors (Speece Cones) in side streams to force mains or gravity sewers

7.3.2 Air Injection

Figure 7-4 is a schematic concept for injection of air or oxygen into a force main. Reports of required air injection rates vary greatly depending on wastewater characteristics, temperature, and pressure in the lines. Very simplistic "rules of thumb" for force mains suggest ranges of 1 to 2 cfm per inch diameter of pipe (U.S. EPA, 1985), but the true demand can be very site specific, require relatively complete characterization of the wastewater and the conveyance system, and potentially include the need to model the collection system.

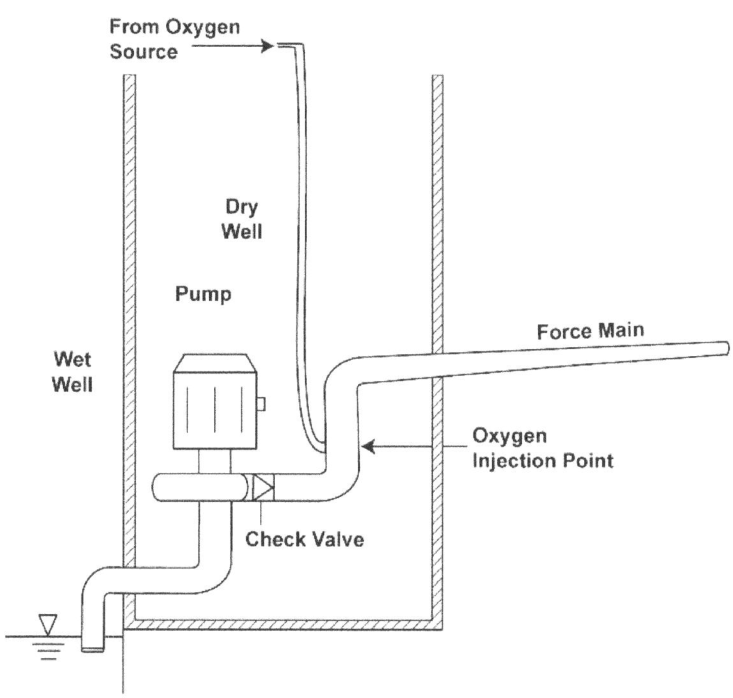

Figure 7-4. Force Main Air Injection.

7.3.2.1 Air-Injection Dosing

Rule-of-thumb recommendations reported by WEF (2004) range from 0.1 to 6.3 cfm per inch of pipe diameter, or on a wastewater flow basis, 0.02 to 0.88 ft^3 of air per gallon of wastewater, with typical reported values between 0.1 and 0.3 ft^3 per gallon.

Soluducha (1989) reported several relatively successful attempts to use air injection in force mains in northern Virginia. In one instance, 25 cfm was injected into a 3-mgd force main with a 2.5-hour detention time, dropping dissolved sulfide from 2.2 down to 0.1 mg/L; this equates to only 0.01 ft^3 per gallon. A second case required 78 cfm in a 4.5-mgd force main to get similar results; this equates to only 0.03 ft^3 per gallon. Soluducha (1989) also reported intermittent maintenance problems with air release valves because of grease buildup plugging the release mechanism actuator of the valves. Maintaining the pressure release valves was reported to be maintenance intensive and problematic.

The air injection approach used a direct stub into a pipe without a diffuser. Air injection rates were calculated using the following equation:

$$V = (mgd \times 8.34 \times Co)/(0.232 \times 1{,}440 \times 0.075) \qquad (7.3.2.1\text{-}1)$$

where:

V = volume of injection air (scfm)

mgd = pumping discharge rate (millions of gallons per day)

8.34 = mass density of water (lbm/gal)

0.232 = mass fraction of oxygen in air

1,440 = unit conversion (min/day)

0.075 = mass density of air (lbm/ft^2)

Co was determined by using the force main residence time at typical flow and an estimated oxygen uptake rate of 10 mg/L per hour. This resulted in 25 scfm, which in this case is sufficient to drop the dissolved sulfides to 0.1 ppm. It should also be noted that this approach assumes 100% oxygen transfer, which may or may not have been achieved in this particular force main application. The force main pressure in this application was reported to have been approximately 85 psi, which may have resulted in a transfer efficiency of nearly 100%.

7.3.3 Venturi Aspirators

Venturi aspirators create a high-velocity flow through a nozzle of decreasing diameter, which creates a suction pressure that can be used to aspirate air into a flowing pipeline. The turbulence in the line mixes the air and water. Fong (1992) reported on full-scale field research in southern California using a Venturi aspirator to inject oxygen in the side stream of a gravity sewer line. Approximately 10% of the flow was side-streamed using a pumping system through a Venturi. Figure 7-5 is a schematic of this type of system.

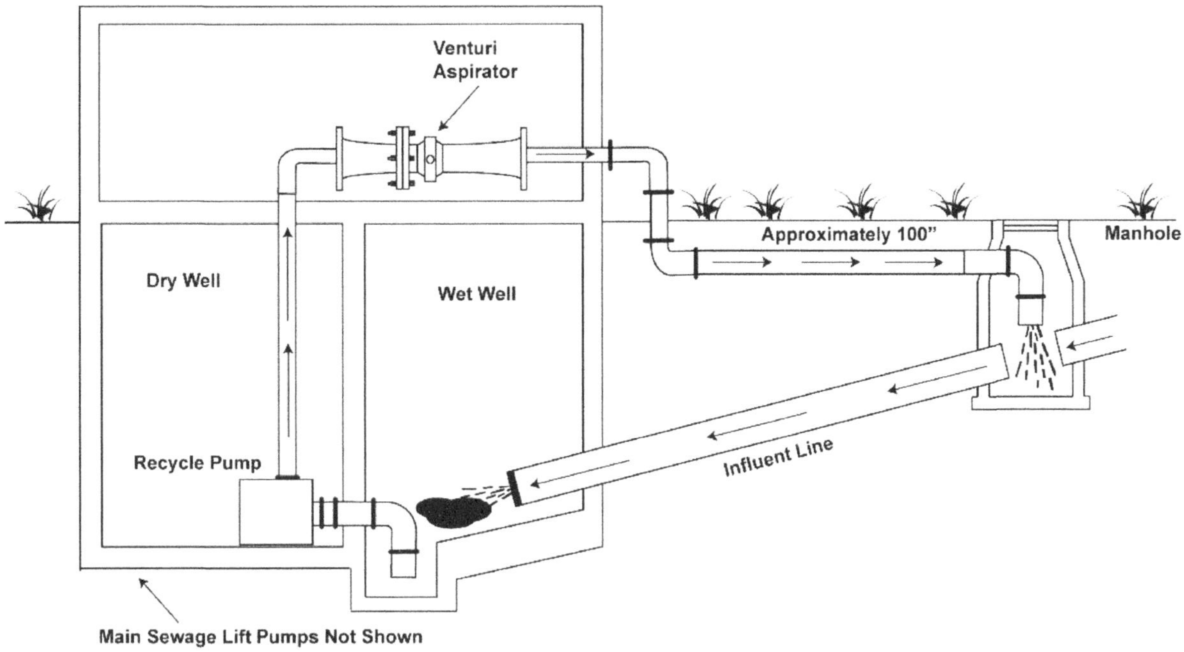

Figure 7-5. Venturi-Style Recycle System.

The optimum (cost-effective) oxygen transfer conditions were found when the system was operated under the following criteria:

- 14 seconds' contact time in the recycle piping
- 70 psi from the side stream pump
- A gas-to-liquid air ratio of 0.08 on a volumetric basis
- Velocity of 14 ft per second in the Venturi

With these settings and under these conditions, 80% oxygen transfer efficiency was achieved. This approach reduced the liquid phase sulfides by 90% and the gas phase H_2S by 70%.

O'Brien et al. (1994) reported the use of a similar system to side stream and recycle a portion of a septage storage tank flow in order to reduce odors from the septage before it was released into the collection system. Two 25-hp pumps were used to recycle 1,000 gpm back into a 100,000-gpd septage-receiving facility providing a 25- to 40-minute recycle time on the reaerated flow. This was done using a vendor-supplied system called a Hydro-Vac. This approach reportedly allowed the facility to stop using $2,000 worth of liquid phase chemicals a day. A similar system is also produced under the supplier name HydroFlo.

7.3.4 Air Lift Pumps

Airlift pumps have reportedly been used to raise oxygen levels at lift pump stations (U.S. EPA, 1985). However, no recent published literature on this topic was identified.

7.3.5 U-tube Aeration

U-tubes have reportedly been used for both air and oxygen injection into gravity sewer lines. The pressure and contact time in the U-tube promotes oxygen transfer. However, no recent published literature about using U-tube aeration was identified.

7.3.6 Pressure Tank Air Injection

Pressurized tanks increase the solubility of air or oxygen. This approach uses a side stream of the flow that is oxygenated under pressure and injected back into the main flow stream. However, no recent published literature about using pressure tank injection was identified.

7.3.7 Hydraulic Fall Injection

Oxygenation can be done using an intentional hydraulic fall when transitioning from a force main into a gravity line. This helps downstream in the gravity line but can release large amounts of H_2S gas at the point of turbulence. However, no recent published literature about using this approach was identified.

7.3.8 Downflow Bubble Contactors

In this case, air or oxygen is injected into the top of a downflowing cone that widens as the flow goes downward. The tendency for the bubbles to be carried downward with the flow is resisted by the bubbles' natural buoyancy, which tries to keep them moving upward. The long bubble contact time provided in the cone promotes high oxygen transfer. Figure 7-6 is a schematic diagram of this type of system.

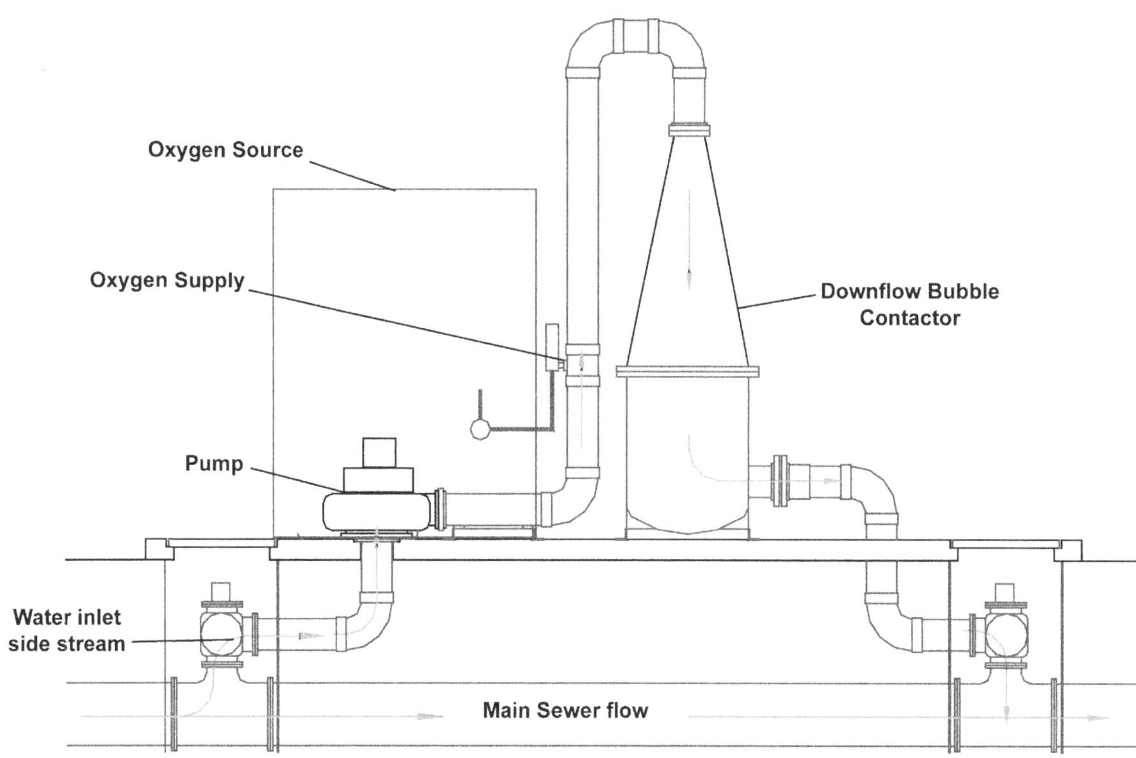

Figure 7-6. Downflow Bubble System Schematic.

Early versions of this device are reported as having been provided by a supplier called Liquid Air and having had up to a 50% transfer efficiency (Melbourne Water, 1989). Speece et al. (2005), McGinnnis et al. (1998), Moore (2003), and Little et al. (2001) report laboratory and field use of Speece Cones to raise collection system oxygen levels as well as dissolved oxygen levels in lakes by treating side streams of flow in a recycle mode. Theoretical oxygen transfer efficiencies as high as 93% were reported. Publications on this type of system in collection systems were limited. "Similar" mixing chamber-style versions of these side stream systems are also provided under the brand name Hydroflo. These systems differ in that the contact chamber relies on mixing baffles rather than downflow bubble contact.

7.3.9 Oxygen Injection

For large-scale use, oxygen can be provided by bulk storage systems of liquid oxygen coupled with finned-style heat exchanger vaporizers. Alternatively, for smaller-scale systems, oxygen can be generated onsite using a pressure swing absorption system.

Prewitt (2000) reported the use of a pressure swing absorption system in Kentucky. In this instance, 90% pure oxygen was generated onsite and injected into a force main. Oxygen demand was based on the assumption of 4.8 lb of oxygen per pound of sulfide; this was reported as effective in controlling sulfide formation.

Advantages of oxygen injection are the following: 1) A preventative residual can be maintained, 2) oxygen may be injected far upstream of the odor problem, 3) ambient air can be used for some applications with pressure swing systems, 4) chemical costs are reduced, and 5) no harmful byproducts are formed. The same mass of oxygen is provided with one-fifth the volume of gas compared to air.

Disadvantages are the following:) Effectiveness may be limited by oxygen solubility, 2) oxygen injection *may* cause gas binding in force mains, 3) it may increase air release valve maintenance costs, and 4) it may not be cost-effective for treating gravity sewers, because injected gas could volatilize out of solution and be wasted. This would depend on how efficiently oxygen is dissolved and distributed.

7.3.9.1 Oxygen Dose

The stoichiometric demand for oxygen depends on which chemical reaction path is assumed. Researchers report a wide variety of chemical reactions promoted by either biological activity or by a non-biologically driven chemical reaction. Which reaction dominates is dependent on pH and the initial ratios of sulfide to oxygen (Melbourne Water, 1989). The most typical and conservative reaction depicted below reacts stoichiometrically at a ratio of 1.9 lb of oxygen per pound of hydrogen sulfide:

$$H_2S + 2O_2 \rightarrow 2H^+ + SO_4^{2-} \qquad (7.3.9.1\text{-}1)$$

Oxygen demand can, however, be driven by more than just the need to react with sulfide, resulting in a higher "effective stoichiometric" demand. The total oxygen requirement demand includes the concentration of dissolved sulfide; the nonsulfide BOD of the wastewater, such as organics and nitrogen-ammonia; and oxygen demand in the slime layers on the collection system walls and in any sedimentation on the bottom of slow-moving collection systems.

Temperature and time in conveyance are important because oxygen is consumed faster at higher temperatures (and oxygen solubility is lower). As such, long-contact-time force mains in relatively high-temperature environments can quickly go anaerobic and begin producing sulfides.

Typical oxygen uptake rates reported for domestic wastewater range from 5 to 15 mg/L/hr (Jordan et al., 2003). Fong (1992) reported oxygen uptake rates of 2 to 15 mg/L/hr. Fong also indicated that the majority of the oxygen reaction is achieved biologically rather than by chemical oxidation paths.

Field trials and collection system modeling were evaluated for a collection system in California by Moody and Riek (1999), who found the typical oxygen dose to be 2.2 lb of oxygen per pound of sulfide removed. Jordan et al. (2003) reported that 4.8 mg/L of oxygen was required to biochemically oxidize each 1 mg/L of sulfide.

Various sources report different levels of background dissolved oxygen required to prevent sulfide formation and buildup. Melbourne Water (1989) reports various researchers' values ranging from as low as 0.5 mg/L to as high as 2.0 mg/L. Tanaka and Takenaka (1994) reported that maintaining 1 mg/L of dissolved oxygen in the terminal point of the force main was sufficient to control sulfide formation.

The applicability of oxygen injection will depend on the BOD exerted over the detention time of the force main and the solubility of oxygen for conditions inside the force main. If more dissolved oxygen is consumed than can be dissolved during the detention time, then oxygen injection may not work without additional injection points.

A variety of equations for calculating oxygen demand and consumption in force mains and collection systems is provided in the literature (e.g., WEF, 1995, 2004; Melbourne Water, 1989; U.S. EPA, 1985).

Various collection system models that can calculate the oxygen demand also exist. These are discussed in Chapter 5.0.

Oxygen solubility can be calculated on the basis of the pressure inside the force main as

$$DO_{max} = P f_O K_H \tag{7.3.9.1-2}$$

where

P = absolute pressure inside the force main (atm)

f_O = fraction of oxygen in the gas to be injected (for air, 0.21 g O_2 / 1 g air)

K_H = Henry's law constant (mg/L/atm) at the wastewater temperature

The Henry's law value for oxygen in water varies as a function of temperature (Figure 7-7).

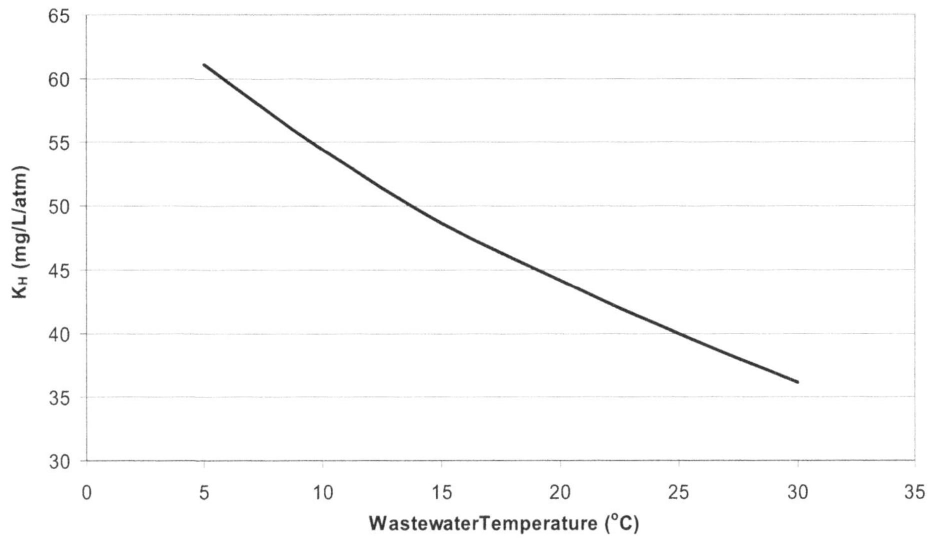

Figure 7-7. Henry's Law for Oxygen as a Function of Water Temperature.

Oxygen solubility should be calculated at the point of injection, the point of lowest pressure inside the force main, and at all air release valves. The amount of oxygen injected should be planned so that as oxygen is consumed along the force main and the oxygen concentration does not exceed saturation at the air release valves and low-pressure points (high-elevation points).

The required gas injection rate can be calculated as

$$\dot{G} = (Q\,DO_{in})/(\eta_{diff} f_O) \qquad (7.3.9.1\text{-}3)$$

where

$\dot{G}$ = gas injection rate (g/s)

Q = wastewater flow rate (m^3/s)

DO_{in} = required dissolved oxygen concentration at the injection point (mg/L)

η_{diff} = diffuser efficiency (gas dissolved/gas injected)

f_O = fraction of oxygen in the gas to be injected (for air, 0.21 g O_2 per 1 g air).

7.3.9.2 Treatment Effectiveness

Although it is consistently reported in the literature (U.S. EPA, 1985; Melbourne Water, 1989; Fong, 1992; WEF, 1995; Jordan et al., 2003; WEF, 2004) that keeping residual oxygen levels above 1 mg/L can control sulfide formation and odors, few data were identified showing the specific effectiveness of this approach on *non*-H_2S odors. Even data on effectively controlling H_2S seemed somewhat limited. This was particularly the case regarding gas phase concentrations of H_2S and other odorants.

7.4 Chemical Oxidation

Aeration or oxygen injection does in fact promote both chemically and biologically mediated oxidation processes; however, there are stronger oxidizing chemicals that are often used to attack odor- and corrosion-causing compounds and destroy them through oxidation-reduction reactions. Strong oxidants can decrease dissolved sulfide concentrations to below the detectable limit. The practicality of different chemical oxidants, however, depends on several factors, including reaction time, chemical cost and availability, safety issues, and ORP. Low ORP (less than -100 mV) tends to indicate that the wastewater is a candidate for chemical oxidation. The discussion of chemical oxidants below includes chlorine-based chemicals, hydrogen peroxide, and potassium permanganate.

7.4.1 Chlorine Gas

7.4.1.1 Technology Background

The first reported chlorine application is described by Melbourne Water (1989), whose text reports that in 1854, chloride of lime was used by the First English Royal Sewage Commission to deodorize London wastewater. Today chlorine is still used in a variety of forms. Chlorine treatment can come from several sources, including chlorine gas, sodium hypochlorite (bleach), and sodium chlorite.

Chlorine gas has been used successfully to oxidize sulfide in sewers, pumping stations, and treatment plants. Although it is used primarily for disinfection, it can also be used to oxidize odorous compounds in the liquid phase. Because of safety concerns regarding workers' and the public's potential exposure to chlorine gas, the use of liquid chlorine and chlorinator systems is typically limited to treatment plants whose facilities are protected and can be monitored and have less potential to impact community safety.

Advantages to using chlorine include the facts that the equipment required is inexpensive and widely available and that chlorine has a long record of effectiveness.

Disadvantages include the following: 1) Reactions *may* form harmful chlorinated organic byproducts, some of which are regulated; 2) chlorine is consumed by ammonia, which may add to the dose requirement; 3) chlorine may inhibit downstream biological treatment; 4) storage and use of sufficient quantities of chlorine can create reporting requirements under risk management plan programs and represent potential safety challenges; and 5) direct injection of chlorine gas may cause downstream fuming and release potentially dangerous chlorine gas (especially if mixing is not rapid and thorough at the injection point).

7.4.1.2 Dose

Chlorine is very reactive with many compounds found in raw domestic wastewater, including hydrogen sulfide. Chlorine and sulfide react as shown by the following:

$$\text{Acidic pH: } HS^- + 4Cl_2 + 4H_2O \rightarrow SO_4^{2-} + 9H^+ + 8Cl^- \quad (7.4.1.2\text{-}1)$$

$$\text{Basic pH: } HS^- + Cl_2 \rightarrow S + H^+ + 2Cl^- \quad (7.4.1.2\text{-}2)$$

Stoichiometrically, 8.57 parts by weight of chlorine are required to oxidize each part of sulfide under acidic pH conditions, whereas 2.14 parts of chlorine per part of sulfide are required for basic conditions. This indicates the sensitivity of the reaction to pH.

However, since chlorine indiscriminately oxidizes any reduced compound in wastewater, competing side reactions also influence the required effective dose.

Field studies have shown that, depending upon pH and other wastewater characteristics, 4 to 15 parts by weight of chlorine are often required for each part of sulfide. These characteristics include BOD, COD, and other oxidation sinks. Lower chlorine demands do some times materialize and may be related to pH as shown in the above equations. Moody and Riek (1999) reported typical demands in the range of 5 to 15. Ten to 15 parts by weight is not, on the basis of U.S. EPA (1985) guidance, unusual, but higher-strength waste may require even higher dosing. For example, Wolf et al. (2000) reported unusually high demands on a high-strength industrial food waste force main. In that study, with a 700 to 1,200 BOD, waste dosing as high as 25:1 was required to sustain effective odor control. This points out the importance of following jar testing with pilot testing to determine the true dosage requirement for a given wastewater and wastewater collection system.

7.4.1.3 Equipment and Handling

Chlorine gas is shipped and stored in tanks containing liquid chlorine under pressure. The gas is drawn off the tank headspace and mixed into a side stream with a chlorinator before being injected into the wastewater. Regulators and diffusers are also required for chlorine injection. Chlorine gas is poisonous and requires special precautions for storage, including containment and scrubbing facilities in case of a leak. Effective mixing is also reported to be very important in full-scale applications (U.S. EPA, 1985; WEF, 2004).

7.4.2 Sodium Hypochlorite

7.4.2.1 Technology Background

Sodium hypochlorite (NaOCl) is the chemical commonly found in bleach home-cleaning products. In typical odor control applications, industrial strength hypochlorite— approximately triple the strength of home-cleaning products—is used.

Hypochlorite is essentially 18-20% caustic with chlorine injected into it, resulting in a 10-15% solution of chlorine stabilized in caustic as hypochlorite. The high-pH caustic stabilizes the chlorine and acts as the carrier. Hypochlorite performs comparably to chlorine gas but is preferable for safety reasons related to potential chlorine gas release.

Chlorine gas tends to acidify any solution it is added to, whereas hypochlorite tends to make solutions more basic, owing to the presence of the caustic carrier solution. Generally, hypochlorite is much safer to use and is preferred over chlorine gas for collection system applications.

Its advantages are that the equipment required is inexpensive and widely available and that it has a long record of effectiveness.

Its disadvantages are that 1) reactions may form toxic and carcinogenic byproducts, some of which are regulated; 2) it may inhibit downstream biological treatment; and 3) it degrades slowly, thus complicating long-term storage.

7.4.2.2 Dose

Hypochlorite is a strong oxidizing agent. It reacts stoichiometrically with hydrogen sulfide in wastewater at a theoretical ratio of 8.8 lb of sodium hypochlorite per 1 lb of hydrogen sulfide:

$$4NaOCl + H_2S \rightarrow 4NaCl + H_2SO_4 \quad (7.4.2.2\text{-}1)$$

It is, however, also sensitive to pH, with stoichiometric demand rates similar to those of chlorine gas injection under acidic conditions.

Since hypochlorite reacts indiscriminately with other wastewater constituents, the required effective dose is often higher than stoichiometric. In practice, the required dosage is often reported to be in the range of 10 to 15 lb of sodium hypochlorite per 1 lb hydrogen sulfide

(U.S. EPA, 1985). However, field experience indicates that significantly lower or higher values can also be observed. Jar testing specific wastewaters is therefore recommended to determine true demand.

7.4.2.3 Equipment and Handling

Equipment required for applying hypochlorite includes a polyethylene or a fiberglass tank, metering pumps, controls, and instrumentation. The equipment is relatively simple to design, operate, and maintain. Personal protection equipment, including gloves, goggles, eyewashes, and safety showers for emergency use, should be included in all chlorine applications, including hypochlorite systems. Under normal conditions, sodium hypochlorite, unlike chlorine gas, is not potentially dangerous to surrounding inhabitants. However, if it were accidentally mixed with an acid, the resulting pH shift could liberate chlorine gas.

7.4.2.4 Treatment Effectiveness

Generally, chlorine is reported to be very effective at reducing H_2S, although the effect can be limited with distance downstream of the dosing point. Non-H_2S odors are also generally reported as being controlled (Choudray, 2000), although hard data on specific non-H_2S odorants was seldom reported.

"Shock dosing" has been attempted using hypochlorite but with short-lasting results. Hamilton et al. (1992) reported that hypochlorite shocking at up to 115 mg/L for as long as 60 minutes on a 4-mile-long collection system achieved a downstream residual of 3 mg/L and did impact dissolved sulfide during the shocking, but within 35 minutes of dosing being stopped, concentrations were back to normal at the downstream end of the reach.

Lamb Technical Services (1998) reported attempts to use hypochlorite shocking in Arizona, where hypochlorite dosing was maintained on a two-mile reach. In this instance sulfide also dropped during the shock dosing, but once dosing was stopped the system rapidly returned to previous liquid and gas phase H_2S levels.

Waltrip and Snyder (1985) reported on continuous dosing evaluations of the force main injection of chlorine in Virginia. Although the information provided is insufficient to determine the dose rate that was used, chlorine was reportedly dosed in a force main carrying 12 to 20 mg/L of dissolved sulfide. Dosing effectively elevated the ORP to +200 and removed the sulfide, but within 1 to 2 miles downstream of the application point, sulfide began to increase again, and within 5 miles sulfide was back up to the initial background levels of 15 to 18 mg/L. This indicates that chlorine can be effective but only for relatively limited periods and that intermediate dosing stations may be needed.

Although chlorine control of odors and corrosion is well established in the literature, few data were identified showing the specific effectiveness of this approach on non-H_2S odors. Even data on H_2S control effectiveness were limited. This was particularly the case with regard to gas phase concentrations of H_2S and other odorants.

7.4.3 Sodium Chlorite

Sodium chlorite ($NaClO_2$), which is produced by Basic Chemical (formerly Vulcan) as a 25% solution, is an alternative source to chlorine sometimes used in sludge applications but not reported in the literature for collection systems. At the time of this writing, sodium chlorite is distributed in the U.S. by Siemens Corporation under the product name VX-456. It reportedly reacts with H_2S as follows:

$$2H_2S + NaClO_2 \rightarrow 2S + 2H_2O + NaCl \quad (7.4.3\text{-}1)$$

Chlorite is used much less often than hypochlorite because it costs more and there is less knowledge of its application. It reportedly reacts with sulfide selectively, so it may have a longer residual than other forms of chlorine. A chemical supplier reported that 4 to 6 lb of $NaClO_2$ per pound of sulfide removed was sufficient to decrease sulfide concentrations to below detectable levels (Rio Linda, 1996).

7.4.4 Permanganate

Sodium and potassium permanganate are strong oxidants that can be used to remove sulfide and other odor- and corrosion-causing compounds. Sodium permanganate is more soluble, whereas potassium permanganate is less expensive, but both are more expensive than other chemicals. Permanganate sometimes is used to treat sludge because it reacts preferentially with sulfide and can be cost effective in this type of application. It is not generally the most cost effective approach on the full flow of a typical wastewater plant collection system.

Permanganate is available in both concentrated liquid and crystal forms. It is most often used as a crystal to make a 3-6% solution. The solution is then blended with the waste stream. In crystal form the concentration as permanganate is high—97%. Also available is a liquid product at approximately 12.5% concentration that typically comes in drum-size carboys or drums for small applications.

It has the following advantages: 1) Solids formed from reaction (manganese dioxide) may have a beneficial effect on downstream flocculation in treatment plants; 2) permanganate can be used to treat sludge; and 3) anecdotal field experience suggests that dewatering equipment will achieve higher dry solids. Permanganate's disadvantages are that it is expensive and that it does not maintain a residual.

7.4.4.1 Dose

Potassium permanganate ($KMnO_4$) is a strong chemical oxidizing agent that reacts with hydrogen sulfide according to the following equations:

$$\text{Acidic pH: } 3H_2S + 2KMnO_4 \rightarrow 3S + 2H_2O + 2KOH + 2MnO_2 \quad (7.4.4.1\text{-}1)$$

$$\text{Basic pH: } 3H_2S + 8KMnO_4 \rightarrow 3K_2SO_4 + 2H_2O + 2KOH + 8MnO_2 \quad (7.4.4.1\text{-}2)$$

In practice, several reactions ranging between these two may take place to produce elemental sulfur, sulfate, thionates, di-thionates, and manganese sulfide depending upon local wastewater chemistry. Field studies have indicated that between six and seven parts of $KMnO_4$ are required for each part of sulfide. A major vendor reports that for sludge streams typical dosing can run as high as 100 mg/L. This often equates to the demand for approximately 7 lb of permanganate per 1 lb of sulfide.

Permanganate reportedly reacts preferentially (first) with sulfide. However, permanganate, like chlorine or peroxide, can be considered an indiscriminate oxidant if sufficient contact time and exposure is provided. As such, dosing requirements can depend on how it is applied.

Very little experience was identified where permanganate is used in collection systems. This appears due primarily to its relatively high cost and complexity of the mixing and delivery systems.

As such, permanganate odor control appears to be substantially limited to side streams in wastewater treatment plants rather than to an entire treatment plant's flow (Lloyd, 1997).

7.4.4.2 Equipment and Handling

Potassium permanganate is typically available and used in dry form and must be mixed with water to an approximately 3-6% solution (with a density of 8.7 lb/gallon) before being used. Mixing with water at temperatures below 40°F may be problematic because of its low solubility; accordingly, heating dilution water may be required. Sodium permanganate is more soluble and is sold in a 12-20% solution.

The equipment required for a typical potassium permanganate application consists of a dry feeder, mixing tank with mixer, metering pumps, control system, valves, and piping. Potassium permanganate typically is used in smaller applications because its cost makes it prohibitively expensive to treat large flows, such as the entire plant flow, on a continuous basis. It is most often used on liquid sludge stream applications.

When kept dry and cool, potassium permanganate is relatively stable. However, when contaminated with organics or acids, it can become unstable and decompose, causing potentially explosive conditions. Thus, sodium permanganate that is stored and transported in solution may be safer in this respect. Appropriate worker safety precautions must be followed when using the material, including use of rubber gloves, goggles, and a particulate breathing mask. Eyewash and shower stations and other safety equipment should be available.

7.4.5 Hydrogen Peroxide

Hydrogen peroxide (H_2O_2) is a strong oxidizer that reacts with most reduced compounds in wastewater, including sulfide. It is typically available in bulk concentrations of 35-50%. It has the advantage of having harmless reaction byproducts, and the breakdown of excess H_2O_2 increases dissolved oxygen in wastewater; a disadvantage is that an H_2O_2 residual is very difficult to maintain (though reaction adds residual O_2).

7.4.5.1 Dose

Hydrogen peroxide oxidizes hydrogen sulfide to colloidal elemental sulfur (Melbourne Water, 1989). Under typical wastewater pH ranges, reactions proceed according to the following equation:

$$pH < 8.5 \quad H_2S + H_2O_2 \rightarrow S + 2H_2O \qquad (7.4.5.1\text{-}1)$$

Field studies and bench-scale test reports indicate a required dose of one to four parts by weight of H_2O_2 per one part sulfide, depending on target sulfide concentration (Van Durme et al., 1987; Waltrip and Snyder, 1985). Larger doses are reportedly needed as the target sulfide concentration approaches zero. For example, a larger H_2O_2/sulfide ratio is needed to achieve a dissolved sulfide concentration of 0.1 mg/L than is needed to achieve a dissolved sulfide concentration of 0.5 mg/L (Waltrip and Snyder, 1985). Because of the fourfold dose range, bench-scale tests on a wastewater sample from the source to be treated are recommended in order to verify the appropriate dose. A field test reported by Melbourne Water (1989) required a peroxide dose up to 40 mg/L for a 5-mile-long pressure main in order to reduce end-of-pipe dissolved sulfides to near zero.

The reaction between hydrogen peroxide and sulfide is relatively slow (reactions are very nearly complete in 30 minutes) when compared to those of other oxidants, which take seconds to minutes. U.S. EPA (1985) reports that 90% of the reaction occurs within 10 to 15 minutes and that reaction is essentially complete within 20 to 30 minutes. Therefore, adequate mixing and contact time must be provided. Melbourne Water (1989) appears to be slightly more conservative and reports up to 60 minutes being required for complete reaction in wastewater. Typical designs shoot for approximately 30 minutes' upstream contact time with good mixing. Melbourne Water (1989) also reports that peroxide reacts with mercaptans, disulfides, and other organic sulfides under acidic conditions in the presence of soluble iron catalyst.

Peroxide-Enhancing Product Ghosn and Cox (2006) report on a relatively new catalytic product that enhances the reaction time of peroxide and other oxidants. In their study, contact times were reduced to 3-5 minutes using the VTX-1 catalyst. This catalyst is reported to be a proprietary organic based metal (iron) complex that enhances the reactivity of the oxidant.

Peroxide Derivative Product Kotonarou et al. (1991) also report on an alternative salt derivative to peroxide called peroxymonosulfate ($HOOSO_3^-$). They report this as more reactive than peroxide but point out that at present there is no economic advantage to this approach owing to the cost of currently available salt products.

Solid Tabular Product Sodium percarbonate is distributed under the brand name Cellinite, which reportedly includes both a time released source of peroxide and that is spiked with bacteriological supplements. This product has been used in small lift stations and in animal waste lagoons for manure treatment (Turo et al., 1998).

Peroxide Regenerated Iron-Sulfide Control Peroxide dosing in combination with iron salts is also reported in the literature under the proprietary name Peroxide Regenerated Iron Sulfide Control. This approach integrates the use of iron salts and peroxide. The combination involves addition of an iron salt in the upper reaches of a collection system followed by injection of peroxide at intermittent downstream locations. The peroxide is used to regenerate the precipitated iron (FeS) recreating ferric and ferrous iron along with colloidal sulfur. The regenerated iron provides subsequent sulfide control.

This approach was reported as beneficial by Walton et al. (2003) in Orange County, California, where reactivation of iron salts at the treatment plant was also reported as beneficial. This report indicates that headspace H_2S in a collection system was reduced from 700 ppm to 20-125 ppm (80-95% reduction), and that 85-90% of the iron added in the collection system could be regenerated at the treatment plant, thereby enhancing both odor control at the plant and settling in the primary clarifiers.

US Peroxide (2001) reported on work at two trunk lines in Orange County, California, and indicated that adding ferrous chloride in upper reaches followed by peroxide at intermediate locations (a 6-hour contact time) and just in front of the wastewater treatment plant dropped liquid phase sulfides from 9 to 12 mg/L down to below 0.5 mg/L.

7.4.5.2 Equipment and Handling

The type of peroxide storage and feed equipment required depends upon the dosage rate and the concentration of peroxide used. The maintenance and repair of hydrogen peroxide systems requires special training and procedures. Hydrogen peroxide is very reactive with organic materials, including grease and oil. When systems are installed or repaired, all wetted surfaces must be specially cleaned and prepared beforehand through a process called passivation. This consists of rigorous cleaning with a sequence of detergent, water, nitric acid, and water to avoid possibly dangerous reactions with contaminants. For this reason, hydrogen peroxide feed systems are sometimes contract operated by suppliers.

Worker safety issues must be addressed when using hydrogen peroxide. Gloves, goggles, and protective clothing should be worn when handling hydrogen peroxide solutions and contaminated equipment because skin, eyes, and other sensitive body parts can be damaged by exposure. Eyewashes and safety showers must be provided for workers. Adequate water supply should also be provided to dilute any spills.

7.4.5.3 Treatment Effectiveness

Peroxide has been reported as effective on both H_2S- and organic-based odor-causing compounds. Devai and Delaune (2002) reported that peroxide was effective in reducing both H_2S and organic compounds such as dimethyl sulfide, methanethiol (methyl mercaptan), and carbon disulfide in a sludge stream when tested on sewage sludge samples.

Moody and Riek (1999) reported the use of peroxide alone at 3.5 lb of peroxide per 1 lb of sulfide in a force main. This dosing level prevented sulfide from rising from 0.3 mg/L to 10 mg/L, observed without peroxide, and maintained the end-of-pipe sulfides to below a target of 0.5 mg/L.

Jefferson et al. (2002) reported on peroxide dosing to condition sewage upstream of a wastewater plant. Dosing of 1.8 mol of peroxide per 1 mol of sulfide at 30 minutes' contact time resulted in approximately a 64% reduction in dissolved sulfides. Increasing the peroxide dose up to 7.4 mol of peroxide per 1 mol of sulfide with 30 minutes' contact time resulted in an 86% removal of liquid phase sulfides. Increasing contact times up to as much as 145 minutes increased the control only up to approximately 88%.

In yet another instance, Nowak et al. (2004) reported using a combination of aeration and peroxide on septage before it was released to a collection system. This reduced by 90% overall odors as measured by D/T; however, required dose rates and reductions of specific odor-causing compounds were not reported.

Schmidt et al. (2000) reported on work studying dosing control methods to improve the efficiency of peroxide at the entrance of a treatment plant in California. The most effective feedback control was based on both flow rate pacing and headspace H_2S levels downstream, where the collection system flow hits the primary clarifiers. Without increasing overall chemical dosing, the headspace H_2S was reduced by 67% by improving the feedback control approach to dosing.

Fong (1986) reported the use of peroxide in large-diameter sewer systems in southern California. At 5:1 and 6:1 doses of pounds of peroxide to pounds of sulfide, the systems achieved 90% liquid phase sulfide control but for only 30 to 60 minutes. Gas phase H_2S was reduced 60 to 100%, and "odors" were reduced 60-90%. At dissolved sulfide levels below 1 mg/L the peroxide reaction is reported as very slow, requiring higher dosing rates. Example installations showed a wide range in required dosing from 2.4:1 at a force main discharge for control to 18:1 in a gravity sewer for prevention. This points out that peroxide dosing to react with existing sulfide may take lower dosing rates than peroxide dosing to control downstream sulfide formation.

Van Durme and Berkenpas (1989) reported on work in an Arizona collection system where 3.3 kg of peroxide per 1 kg sulfide reduced odors but did not meet the 0.5 mg/L H_2S liquid phase goal.

7.5 Sulfur Precipitation

Various metal salts can combine chemically with dissolved sulfide to form relatively insoluble metal sulfide that precipitate from the wastewater, thus preventing the release of H_2S gas. Metal sulfide precipitates exist as soft black or reddish-brown flocs that usually do not settle in the collection system but can be removed at treatment plants. In fact, more-restrictive industrial source control programs have been credited with a tendency for increasing levels of sulfides over the years because of the associated lowering of metals in the wastewater (Padival, 1993). Effective metals might include iron, zinc, lead, chromium, cadmium, nickel, and copper, but iron is most typically used today owing to availability of the chemicals and the fact that other metals have become more restrictive in terms of allowable plant effluent and biosolids impacts. This is a potential concern because heavy metal concentrations could be regulated and make the solids difficult to dispose of.

Sulfur precipitation has several advantages, including the following: 1) long residuals can be maintained to precipitate sulfide as it is generated, which can result in sulfide control over very long collection system reaches; 2) iron salts can be used to treat sludge; 3) reaction byproducts are innocuous; and 4) precipitates are beneficial to downstream treatment processes.

The following are its disadvantages: 1) Precipitates can dissociate at pH less than 6.5, allowing sulfide to release back into the wastewater (U.S. EPA, 1985); 2) at typical wastewater pH, dissolved sulfide cannot be decreased to much lower than 0.2 to 0.5 mg/L; 3) anecdotal evidence suggests that precipitates form "clinkers" when iron-dosed biosolids are incinerated, and therefore this potential should be considered where sludge cake is incinerated; 4) iron salts can form a film on pipe walls, instrument sensors, and ultraviolet disinfection equipment; and 5) these precipitates can substantially increase sludge production at the treatment plan.

7.5.1 Reaction Mechanisms

Ferrous iron salts such as ferric chloride ($FeCl_2$) and ferrous sulfate ($FeSO_4$) are commonly available for sulfur precipitation. Ferrous iron reacts with each sulfide species as shown below:

$$Fe^{2+} + S^{2-} \rightarrow FeS \quad (7.5.1\text{-}1)$$

$$Fe^{2+} + HS^{-} \rightarrow FeS + H^{+} \quad (7.5.1\text{-}2)$$

$$Fe^{2+} + H_2S \rightarrow FeS + 2H^{+} \quad (7.5.1\text{-}3)$$

These reactions indicate a stoichiometric dose of 3.8 lb of $FeCl_2$ per 1 lb of sulfide. However, effective doses for reducing the sulfide concentration to below 1.0 mg/L are reported from 4 to 12 lb of $FeCl_2$ per 1 lb of sulfide removed, depending on the target sulfide concentration (Slezak et al., 1994; Mohleji et al., 1992).

Ferric iron reacts similarly to ferrous iron:

$$2Fe^{3+} + S2^{-} \rightarrow 2Fe^{2+} + S \quad (7.5.1\text{-}4)$$

$$2Fe^{3+} + 3HS^{-} \rightarrow Fe_2S_3 + 3H^{+} \quad (7.5.1\text{-}5)$$

These reactions indicate a "stoichiometric" dose of 3.3 to 4.9 lb $FeCl_3$ per 1 lb of sulfide. However, field and laboratory experiments indicate that the typical required dose to remove sulfide in domestic wastewater is between 3 and 7 lb $FeCl_3$ per 1 lb of sulfide removed (Jameel, 1989; Van Durme and Berkenpas, 1987).

Ferric iron solutions have a lower pH than does ferrous iron. Care should be taken that the wastewater is sufficiently buffered so that its pH is not lowered by ferric salts.

Ferrous and ferric iron in combination are often used for sulfide precipitation and odor control, because the combination has a synergistic effect such that the overall dosage is reduced. The following equations show these reactions:

$$Fe^{2+} + S^{-2} \rightarrow FeS \quad (7.5.1\text{-}6)$$

$$2Fe^{3+} + S^{-2} \rightarrow 2Fe^{+2} + S^{o} \quad (7.5.1\text{-}7)$$

$$Fe^{2+} + 2Fe^{3+} + 4HS^{-} \rightarrow Fe_3S_4(s) + 4H^{+} \quad (7.5.1\text{-}8)$$

$$Fe^{2+} + HS^{-} \rightarrow FeS + H^{+} \quad (7.5.1\text{-}9)$$

$FeSO_4$ and $FeCl_3$ in combination have been demonstrated to be more effective than either chemical alone. Field experience with reductions down to 0.2 mg/L dissolved H_2S have been reported. Typically, one part of ferrous $FeSO_4$ is used in combination with two parts ferric $FeCl_3$. The limitation of this approach is the requirement for separate chemical feed equipment for each chemical.

Another promising combination is iron salts followed by hydrogen peroxide. In a process sold as Peroxide Regenerated Iron Sulfide Control by US Peroxide, iron salts are used to complex with sulfide regenerated by means of downstream peroxide dosing. This chemical combination is discussed also in Section 7.4.5.

7.5.2 Dosing Requirements

Won (1988) reported on work done in large-diameter sewers in Los Angeles County, California, and indicates that the effective stoichiometric iron dose can be very dependent on the background concentrations of sulfide. For lower concentrations, higher ratios may be required. This paper reported that for wastewater with already low sulfide concentrations below 1 mg/L, up to 100 lb of iron might be needed for each additional pound of sulfide controlled whereas if the background sulfides were above 4 mg/L, the effective dose ratio was closer to 7:1.

Mohleiji et al. (1992) reported on ferrous chloride dosing into sewers in California. In this case ferrous dosing of less than 15 mg/L dropped dissolved sulfides to below the detection limits but left residual gas phase H_2S. Dosing of 20 mg/L was sufficient to drop both liquid phase and gas phase levels to below detection. Dosing in the range of 25-30 mg/L also lowered the gas phase mercaptan and dimethyl sulfide levels to below the detection limit. This is the only case found in the literature that reported iron salt impacts on non-hydrogen sulfide odors.

Richard et al. (1993) reported on the use of ferrous chloride in North Dakota collection systems where 8 lb of iron per 1 lb of sulfide was sufficient to reduce the dissolved sulfides to a target level of 0.5 mg/L. At this dosage, suspended solids increased 37% at the downstream treatment plant and the pH dropped by 0.2 units.

Slezak et al. (1994) reported on the use of "pickle liquor," an industrial waste acidic iron solution byproduct created by an acid-washing process in the steel industry. In this instance, the authority wanted to evaluate both the effectiveness of this low-cost iron source and the impact on the plant's effluent streams, including liquid and biosolids waste. Dosing of 10 to 15 mg/L iron was found to be sufficient to keep sulfides below 0.5 mg/L. This demonstration study indicated that this approach might be feasible as long as pickle liquor-dosing locations in the collection system were controlled, and the effects on the plant were closely monitored. In this instance, it was the iron that was the concern rather than other regulated metals.

7.5.3 Using Iron Coagulated Sludge in Sewers

Several papers reported on attempts to use iron-coagulated sludge from drinking water plants to facilitate sulfide control in the collection system. Edwards et al. (1997) reported on recycling waste iron coagulated sludge in this manner and indicate that when plausible this can be a cost effective control method. They also reported that the effects of this method extended into the treatment plant, resulting in lower H_2S levels in digester gas.

7.5.4 Potassium Ferrate

Sharma (2002) provided a summary for an alternative iron product, potassium ferrate. Potassium ferrate (K_2FeO_4) provides iron (VI), a powerful oxidant that might be used for either disinfection or odor control. At neutral pH, it takes 1.51 K_2FeO_4 per 1 sulfide. Full-scale field application of this oxidant for odor control was not discussed.

7.5.5 Handling and Equipment

The equipment required for dosing metal salts is similar to that in other liquid chemical-handling systems. However, iron salt solutions are corrosive and hazardous and have a low pH, so special precautions must be taken in handling and storing iron salt solutions. The chemical delivery systems typically require a polyethylene- or fiberglass-reinforced plastic tank, metering pumps, associated valves and piping, adequate spill control, and safety equipment.

7.5.6 Treatment Effectiveness

Some literature suggests that iron control has a limit below which liquid phase sulfides cannot be controlled by iron salts alone. The achievable levels reported vary greatly. Nielsen et al. (2005a) reports that iron salts may not be able to lower the liquid phase sulfide below 0.2 mg/L. This same paper reports that iron reactions can be dependent on whether the iron is added under anaerobic or aerobic conditions and that aerobic conditions produced lower results. Sercombe (1995) reports on work by others that indicates iron salts could not drop the liquid phase sulfide below 4 mg/L. Bielefeldt et al. (2002) reported on both lab-scale and full-scale testing results for ferric chloride in wastewater just upstream of a treatment plant and on the sludge. This paper suggests that control down to 0.5 mg/L of dissolved sulfide is possible with ferric iron salts.

Jameel (1989) reports successful treatment down to 0.5 mg/L dissolved sulfides on collection systems in Arizona. Required dosing used 6 gallons per hour of 40% ferrous chloride dosed into 9.5 mgd, an equivalent of approximately 23 mg/L. This was also reported as 4.3 lb of ferrous chloride per lb of sulfide removed.

Dewaters et al. (1999) reported on work in waste treatment lagoons that shows the effectiveness of the presence of oxygen enhancing the performance of iron salts, indicating that lower dissolved sulfide levels can be obtained if any oxygen is present. Without oxygen, iron's effective limit was 1 mg/L; with it, the iron sulfide precipitation dropped levels even lower.

Padival et al. (1995) reports iron salt dosing on a 44-km trunk sewer reach in Los Angeles. The impact on the upper 27 km was reduction of the average dissolved sulfides by as much as 94%. This required approximately 16 mg/L Fe dosing. A blend of 1.9:1 ferrous to ferric worked best, achieving a 97% reduction. Below the 27-km mark, the impacts appeared to be diminished. The paper suggests this was due to impacts of side streams. This is an effective example of how far-reaching iron control can be downstream of the dosing point.

Bielefldt et al. (2002) reported on both lab-scale-testing and full-scale-testing results for ferric chloride on wastewater just upstream of a treatment plant. Ferric was shown to increase suspended solids, resulting in higher primary sludge blankets that became poorly defined (reportedly from pin floc in the primaries). Ferric was shown to create process impacts, including 50-95% reduction in sulfide in downstream wastewater plant anaerobic digesters. Reduced performance in belt filter press dewatering was also reported. Ferric doses below 4.5 mg/L helped reduce odors without impacting processes. Ferric dosing above 4.5 appeared to negatively impact sludge volume indices. Ferric dosing above 13.5 mg/L impacted belt filter press performance. Dosing at the full scale of as low as 2 mg/L reduced but did not get rid of odors (more was needed). An optimum range appears to be near 5 to 12 mg/L.

Wolstenholme and Jabloner (2004) reported that dosing of 8 mg/L of ferrous chloride reduced gas phase H_2S in a naval station's intermittent force main discharge from an average of 35 ppm to below 1 mg/L. In this instance, waste from naval ships (with high detention times) was being pumped through a force main and impacting the receiving collection system.

Padival (1993) reported on full-scale testing on trunk sewers in California to control dissolved sulfide level. Padival reported that 16 mg/L of blended Fe_3 and Fe_2 at a ratio of 1.9:1 provided a 97% reduction in the dissolved sulfides and a 79% reduction in the headspace H_2S. The blended ferrous and ferric worked better than they did alone. O_2 reportedly made Fe_2 as effective as Fe_3.

7.6 Nitrate Addition

Nitrate salts can be used to prevent the formation of sulfides by providing an alternative electron acceptor for sulfate-reducing bacteria, which preferentially reduce oxygen, nitrate, and sulfate in that order. Therefore, as long as enough oxygen or nitrate is in solution, sulfate conversion to sulfide will be minimized. Nitrate can also be used as an oxygen source to promote the biological oxidation of sulfide. Nitrate is available in both liquid and dry form as sodium nitrate ($NaNO_3$) or, more commonly, as calcium nitrate ($Ca(NO_3)_2$). Both salts are quite soluble, and liquid solutions of between 40 and 60% by weight are usually used. Several products, such as Nutriox and Bioxide, are identified in the literature. Bioxide is a trademark for a calcium nitrate solution currently distributed by US Filter throughout the United States. US Filter also distributes Bioxide-AQ, which includes a slime layer bacteria inhibitor called anthraquinone. This will be discussed in the following section.

Nitrates have been used to control sulfides in collection systems, in sludge, and in treatment lagoons (Hobson and Yang, 2000; Einarsen et al., 2000; Wissenberger et al., 2004; Groleau et al., 2002; Watson et al., 2001). Einarsen et al. (2000) reported on using nitrate mixed into dewatered centrifuge cake to control residual cake odors; H_2S- and mercaptans-related odors were reduced.

Watson et al. (2001) reported success in pulp and paper treatment lagoons where dissolved sulfide levels were reduced from around 15 to 20 mg/L down to below 0.1 mg/L. This process reportedly took one to two weeks. This is considered indicative of the biological process changes that must occur with Bioxide. Collection system changes are also not instantaneous and can reportedly take from hours to days.

Nitrates are also created naturally as the side effect of nitrification. These naturally created nitrates can be used to inhibit sulfide generation. As an example, Delgado et al. (2004) reported on using residual nitrate present in plant effluent to control sulfide formation in long-contact-time force mains conveying reclaimed wastewater. Rodriquez-Gomez et al. (2005) reported similar work, where nitrate dosing was used to control sulfide formation in reclaimed water pipes. Coughenour et al. (2003) reported on work investigating the recycling of centrate discharge from dewatering to a dedicated activated sludge reactor for nitrification creating nitrate. Field trials of nitrate addition at the plant headworks reportedly dropped H_2S odor levels throughout the plant.

Nitrate addition has the following advantages: 1) It can maintain a long residual; 2) it is safe and does not require special handling or storage; and 3) reactions do not form hazardous byproducts. Its disadvantages are that nitrate 1) sometimes causes gas binding of long force mains with inadequate air relief valves and 2) may increase the cost and difficulty of denitrification at the treatment plant.

7.6.1 Dose

Nitrate addition reportedly controls dissolved sulfides by two methods, prevention and control. Prevention is based on nitrate's acting as the alternative electron acceptor and promoting the use of nitrate instead of sulfate by the facultative and obligate anaerobic bacteria, thereby preventing the formation of sulfide. Okabe et al. (2003) reported on work on sewer system biofilms. In their study, adding nitrate forced the sulfate reduction zone deeper into the biofilm and significantly reduced the volumetric sulfate reduction rate. This is considered direct evidence of the impact of nitrate on the activity in the sewer systems biofilm.

In the removal reaction, the nitrate supplies oxygen, which allows biochemical oxidation of sulfide by bacteria such as *Thiobaccillus denitrificans*.

Nitrate addition can reportedly control dissolved sulfide concentrations exceeding 0.5 mg/L or less. The required dose ranges from 4 to 15 lb of calcium nitrate per pound of sulfide prevented (Moody and Riek, 1999; Hamilton et al., 1992; George Butler Associates, Inc., 1999).

A minimum requirement is 7.2 kg of nitrate per kilogram of sulfide prevented (Davis Process, 1988; WEF, 2004). This does not include any demand for reacting with existing sulfides. This requires an additional 2.4 kg of nitrate per pound of sulfide reacted. Total requirement is 7.2 plus 2.4, or 9.6 kg nitrate per 1 kg sulfide controlled.

Perry et al. (1997) reported on work in four different force mains. These force mains required nitrate dosing of 16, 12, 9.6, and 10.4 kg of nitrate per kilogram of sulfide. Data reported generally appear to indicate that nitrate dosing upstream needs to be sufficient to keep residual downstream at the discharge point in the range of 0.4 to 0.8 mg/L in order to be effective at keeping headspace H_2S below 10 ppm. This appeared to equate to approximately 12 mg of nitrate per milligram of sulfide reacted or prevented.

It is sometimes noted that the true dose is dependent on the layout of the collection system, which affects the amount of slime layer that must be impacted, and as a result, dosing requirements based solely on pounds of sulfide prevented or reacted may not be accurate. Thus, trying to determine accurate nitrate-dosing requirements by means of jar testing would likely prove ineffective because the natural slime layer is not present in a jar test. Melbourne Water (1989) reported ratios as high as 20 or 30 lb of nitrate per 1 lb of sulfide. This may be indicative of the impacts of the slime layer on the collection system walls and in sediment in the bottom of slow-moving collection systems. Hamilton et al. (1992) indicated that calcium nitrate dosing requirements were 2.7 gallons per pound of sulfide for large trunk sewers and 1.5 gallons per pound for small lines for typical off-the-shelf Bioxide solutions. This may reflect the impact of different levels of slime layer coatings in terms of both surface area and thickness. Nitrate dosing must be continuous. Batch dosing reportedly does not work (Hobson and Yang, 2000). This is likely because the nitrate is consumed, and sulfate-reducing bacteria quickly reestablish themselves.

Pope (1996) reported that calcium nitrate dosing requires field testing and cannot be done effectively with a jar test. Pope also indicated that calcium nitrate worked better for the force main application than the gravity lines. Vollersten et al. (2005a) estimated during batch jar testing 4.6 g of nitrate required per gram of sulfide. This dose is well below the 10.4:1 ratio highlighted above for both prevention and reaction. This relatively low dose may be a result of jar testing that does not include all the effects that would be seen in the field.

Although a dose rate must be selected in order to design the delivery system, the rate should be varied during operation to avoid waste and excess nitrate loading in the downstream process. One approach, if the necessary monitoring equipment can be installed, is to vary the nitrate dose on the basis of the nitrate concentration entering the treatment plant or at the discharge point of the force main. During startup, the nitrate dose is varied until a minimum nitrate concentration entering the plant headworks or reaching the downstream end of the force main is found that maintains the sulfide concentration below the target. Thereafter, the feed rate is slaved to the sulfide concentration and to the nitrate concentration entering the plant so that the previously determined residual concentration is maintained.

7.6.2 Handling and Equipment

Nitrate salt solutions are neither corrosive nor hazardous and do not require special equipment. Nitrate feed systems include a storage tank with metering pumps and associated valves and piping. Figure 7-8 shows a feed system for US Filter's nitrate brand Bioxide.

Figure 7-8. Bioxide Storage and Delivery System.

7.6.3 Treatment Effectiveness

Hobson and Yang (2000) noted that dissolved sulfide was suppressed in rising mains down to 0.1 mg/L with a corresponding suppression in overall odor. In the same system where sulfide was controlled by iron dosing, non-H_2S odor residual was reported.

Bentzen et al. (1995) reported on a proprietary approach in the United Kingdom to dosing sodium nitrate (Nutriox). Nitrate was dosed upstream of a wastewater plant at a pump station in an attempt to keep excess nitrate from entering the treatment plant downstream. The intent was to keep wastewater treatment plant sulfide loading below 0.5 mg/L and nitrates below 2 mg/L. Dosing reportedly controlled liquid and gas phase H_2S being formed in the force main coming into the plant and downstream in the plant's primary clarifiers. Typical dosing was 319 L per day to the 1,200-m^3/day plant flow (equal to 266 gallons of Nutriox per million gallons of wastewater). This reduced dissolved sulfides from around 4 mg/L to an average of 0.24 mg/L (a 95% reduction). Primary clarifier effluent sulfide levels were also reduced from around 5.4 mg/L to 2 mg/L. This means there was a 63% reduction in the primary clarifier H_2S levels as a side effect of dosing Nutriox in the collection system. Nitrate levels entering the plant were around 2 to 3 mg/L.

Sercome (1994) reported on work in the United Kingdom in a system with many pump stations and force mains. Over a four- to five-day period, sodium nitrate dropped headspace H_2S from 50 to 1 ppm at the discharge of the rising main.

Yang (1999) presented work that suggested that in a side-by-side study on a rising main, both H_2S and overall odor units are reduced by Nutriox (nitrate), but that iron salts affected only H_2S and had high residual non-H_2S odors as measured by odor panel odor units.

Churchill and Elmer (1999) reported on the use of commercial-grade $CaNO_3$ (not Nutriox or Bioxide) at a 1.5% solution to control force H_2S emissions that were coming from a hydraulic drop in a collection system section just downstream of a 21-in. gravity sewer's slow-moving flow that was partially fed by an 8-in. force main with 90 minutes of detention time. Nitrate dosing dropped the gas phase H_2S from and average of 68 ppm to near zero in approximately three days.

Nowak et al. (2004) reported on work in Vienna, Austria, where calcium nitrate was dosed at 10 to 20 mg/L, reducing odor levels peaking as high as 33,000 odor units/m^3 and H_2S levels of only 5 to 10 ppm. The paper suggests that odors were being caused by more than H_2S, although other specific compounds were not defined. At this nitrate dosing, H_2S formation was "prevented," and odor concentrations "declined significantly." A review of data plots suggests that nitrates dropped the background odor units by approximately 85%.

Hagekhalil et al. (1994) reported on the use of a 54% calcium nitrate solution in gravity collection systems in southern California. Nitrate residuals in the range of 0.58 mg/L at the downstream end dropped liquid and gas phase H_2S down from 1.9 mg/L of dissolved liquid sulfide and 91 ppm gas phase to 0.02 mg/L liquid and 2 ppmv gas (an over 98% reduction). This required a calcium nitrate dose of 2.31 to 4.4 gallons nitrate per pound of sulfide removed, depending on the system. This equated to 10.4 to 20 lb nitrate per pound of sulfide.

7.6.4 Downstream Treatment Plant Effects

Bentzen et al. (1995) reported that nitrate dosing in the collection system reduced the soluble BOD reaching the plant by 35%. Also noted at the plant was improved nitrification during the nitrate-dosing period, which was associated with the lower BOD and soluble sulfide load.

Moody and Riek (1999) reported that nitrate dosing several hours upstream of a treatment plant was successful at keeping dissolved sulfide to below 0.5 mg/L. When this occurred, residual nitrate levels at the plant reportedly caused "denitrification in the primary clarifier solids and floatation of the solids." Adjusting the target sulfide concentration upward to 1 mg/L and reducing the residual nitrate hitting the plant diminished the impacts on solids floatation in the primaries.

7.7 pH Adjustment

pH can be an important parameter for most of the liquid phase odor control methods described above, but it can also be used directly to mitigate hydrogen sulfide release. There are two methods of reducing H_2S-related odors using pH adjustment. One is to shift the wastewater pH so that chemical equilibrium favors nonvolatile sulfide species. The other uses a temporary but more drastic pH increase to deactivate the sulfur-reducing bacteria.

7.7.1 pH Stabilization

pH affects the distribution of the three sulfide species—H_2S, HS^-, and S^{-2}—in the solution. An increase in pH drives sulfide to ionic forms (HS^- and S^{-2}) that are nonvolatile. For example, at a pH of 7, roughly 50% of sulfides are present as H_2S, whereas at a pH of 8.5, only 3% are present as H_2S. If wastewater pH drops, the sulfide species ratios shift rapidly toward H_2S, which is volatile. Therefore, increasing wastewater pH can reduce odor problems as long as the higher pH is maintained. Figure 7-9 shows the relative fraction of each sulfide species as function of pH.

Figure 7-9, which was generated on the basis of pk_a values for hydrogen sulfide, shows that at a pH of 9 or above, almost no dissolved sulfide is in the volatile form. Some literature also suggests that slight shifts in pH can begin to inhibit growths of anaerobic organisms that produce sulfides (Melbourne Water, 1989).

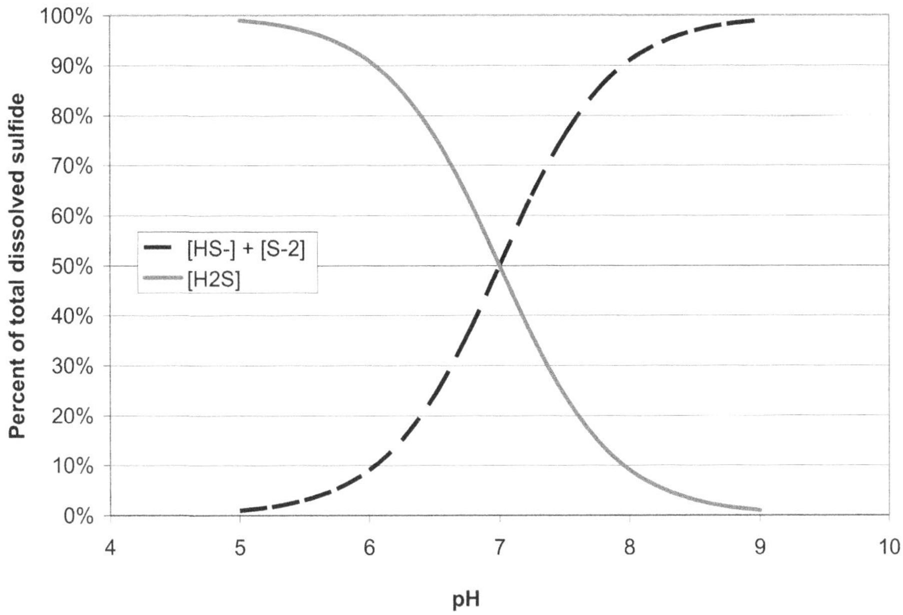

Figure 7-9. Fraction Dissolved Sulfide (Volatile and Nonvolatile).

7.7.1.1 Dose

Magnesium hydroxide ($Mg(OH)_2$) and sodium hydroxide (NaOH) are the two chemicals most commonly used to raise pH. Magnesium hydroxide can raise pH to a maximum of 9 and sodium hydroxide to a pH of 14. Even with overdosing, magnesium hydroxide reportedly does not raise the pH over 9 (Moore, 2003a), reportedly because it is not as soluble and acts as a buffer, whereas caustic is readily soluble and can have a quick effect on pH, driving it up much higher.

Jefferson et al. (2002) reported on work just upstream of a treatment plant. In this instance, magnesium hydroxide worked well to adjust pH up to about 9. Owing to the relatively insolubility of MgOH, overdosing only caused the pH to remain stable at up to 9.5.

The chemical dose will depend on the buffering capacity of the wastewater. To estimate the amount of chemical needed to achieve the target pH, a sample of the wastewater should be jar tested with a concentrated solution of the desired control chemical until the target pH is reached. Lime has also reportedly been used (U.S. EPA, 1985). Dosing experiences for each chemical are outlined below.

Magnesium hydroxide is most commonly known as an acid-stomach remedy and is essentially industrial-grade milk of magnesia. Typically, magnesium hydroxide comes in a 58-60% slurry with a paintlike consistency. It freezes near 32°F and should be stored in stirred or agitated tanks. It reportedly works by pH shifting to keep already-formed sulfides in solution, sulfide binding with magnesium cations, and pH inhibition of the slime layer. It is also reported to have beneficial downstream impacts by providing alkalinity to the treatment plant processes. Dosing can be either direct to the wastewater or as an intermittent spray coating on the crown of gravity collection systems.

Reported rates of liquid phase dosing directly to the wastewater vary greatly. Moore (2003b) reported typical dosing requirements ranging from 25 to 100 gallons per million gallons of wastewater treated. It can either be applied directly to the wastewater flow or sprayed on the crown of a gravity sewer line.

Magnesium hydroxide was tested as a sulfide control chemical in CH2M HILL's (2001) Orange County Sanitation District study. Roughly 700 pounds of magnesium hydroxide per million gallons were dosed as a slurry to a long gravity interceptor. The vapor phase H_2S concentration went from 110 to 21 ppm, a substantial gas phase H_2S reduction. These results are in contrast to manufacturer-recommended doses of between 10 and 100 gallons of 58% solution per million gallons. If possible, jar tests should be conducted to determine the appropriate dose for the wastewater in question. When this method is used, care should be exercised that dilution from side sewers does not lead to a pH drop and downstream odor problem.

7.7.1.2 Equipment and Handling

pH stabilization requires storage and feed equipment similar to those in other liquid feed systems. Magnesium hydroxide is nonhazardous and safe to handle; however, it is applied as a slurry and requires mixing equipment.

7.7.2 Caustic Shock Loading

Odor also can be controlled by shock-dosing the H_2S-producing bacteria living in the slime layer on the sewer wall. Raising the pH of the wastewater to 13.0 or higher for 30 minutes has been demonstrated to inactivate the sulfate-reducing bacteria for periods of 3 to 14 days. Churchill and Elmer (1999) reported on work where shock dosing was used to control odor release from a force main; pH shocking was reported as effective but "short lived."

7.7.2.1 Dose

A Los Angeles field study found that raising the pH in a gravity sewer to 13.2 for 30 minutes decreased the sulfide concentration from 4.5 mg/L to zero on the first day. By the third day after shock loading, the sulfide concentration was up to 0.25 mg/L, and it continued to increase thereafter. The study also described shock loading up to a pH of 11.4, which did not work (Hamilton et al., 1992). The appropriate dose should be determined by titrating a sample of wastewater with the NaOH solution to be used until a pH of 13 is reached. A study by the Los Angeles Sanitation District found that a dose rate of 3,000 lb of 50% NaOH solution (260 gallons) per million gallons of wastewater for 30 minutes was required to obtain odor control by shock loading in their system with their wastewater.

7.7.2.2 Equipment and Handling

Shock dosing is often applied directly from a tanker truck. This method saves the capital cost of storage and feed equipment. Sodium hydroxide is dangerous to handle, and so appropriate equipment and safety measures are required for its use.

7.7.2.3 Treatment Effectiveness

pH Adjustment Using Caustic Khan (1992) reported on sewer corrosion studies done in Southern California using continuous caustic dosing to adjust pH levels. For a jar test on the specific Los Angeles wastewater system, it took 80 mg/L to adjust the pH to 9. In field application, 91 mg/L was actually required. This level of dosing and pH adjustment decreased manhole gas phase H_2S by 71-78%. Manhole odor levels in terms of D/T reduction decreased from 2,800 D/T to 300 D/T.

pH Adjustment Using Magnesium Hydroxide Orange County Sanitation District (2000) reported on dosing Thioguard (a proprietary magnesium hydroxide solution) into trunk sewers in southern California. The required dosing was 96 gallons MgOH per million gallons of wastewater. This reduced headspace H_2S concentrations by 86-98%. No discussion of broad-spectrum odor reduction was provided.

Caustic pH Shocking Crowley et al. (2002) reported on pH shocking on a pump station force main in North Carolina. A 3-mile-long, 30-in.-diameter ductile iron force main discharged into a 9.15-ft drop into a manhole where H_2S odors were released. In this instance, pH shocking up to 12 for 20 minutes or more got rid of H_2S odor emissions but only for 4 days. It took 3 mL per 1,000 mL of 25% caustic added to wastewater to raise the pH to 12. This reduced the dissolved sulfide by only 43% but virtually removed gas phase sulfides that came back in four days. This was reinforced in findings provided by Walton et al. (2003), indicating that partial biofilm recovery (33-55%) occurred within 1 to 2 days, and complete recovery was observed in 3 to 4 days. This was also reported by Hamilton et al. (1992).

Jin (1987) reported results of sodium hydroxide shocking on large-diameter sewers in southern California. This study focused on large-diameter (51-144 in.) gravity sewers in order to determine the economic impacts of this approach to odor and corrosion control. Projected chemical costs from this 1987 study were estimated to be nearly $15 million annually to achieve 95% control of dissolved sulfide. The paper also describes the history of the pH-shocking program and indicates that initial shocking was attempted with lime slurries but that doing so did not prove effective and was problematic because of the settling of the slurry solids, which caused sewer-cleaning problems. The overall program was changed in favor of using caustic solutions delivered from trucks.

The literature reports that in most instances, caustic shocking would have to be repeated every one to two weeks. This research recommended three treatments a week at 12.6 pH to achieve 90% control of dissolved sulfide. Raising the pH above 13 showed further reduction, but recovery was still complete within two weeks. Gas phase reductions were significantly less. For instance, if dissolve sulfides were reduced 74%, then gas was reduced 48%. It was also noted that caustic treatment did not affect the crown pH on the concrete surface. This paper (Jin, 1987) reported a wide range in recovery of 38-100% slime layer recovery within seven days and complete recovery in two weeks. Dissolved sulfide reduction plots in this study indicate the levels can return within three days. Figure 7-10 is a copy of the testing reported for smaller-diameter (approximately 24 in.) sewers. Similar results were reported for larger-diameter ones.

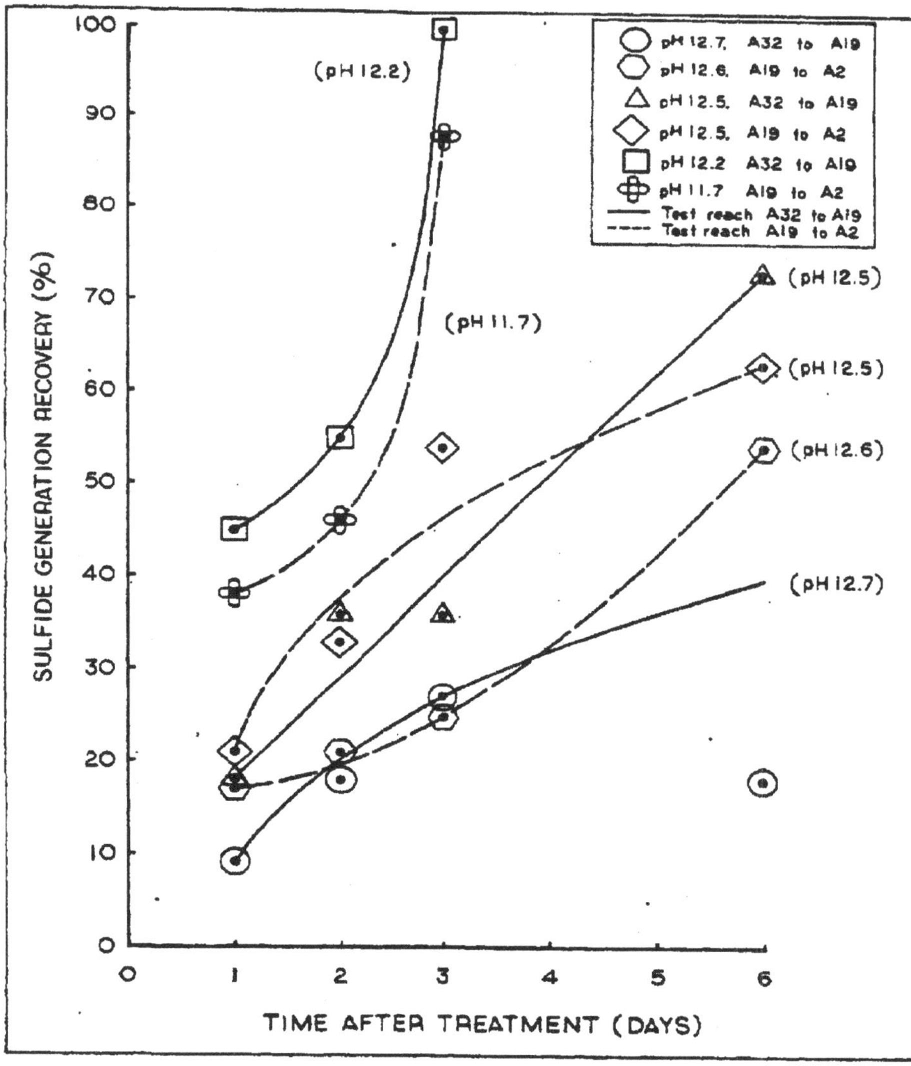

Figure 7-10. Sulfide Generation Recovery in Joint Outfall "A" with Caustic Treatment.

7.7.2.4 Downstream Treatment Plant Effects

Melbourne Water (1989) indicates that wastewater with a pH level above 9.0 entering a downstream treatment plant can be harmful to biological processes, such as an activated sludge or fixed film treatment system. Dao (2004) indicated that periodic shock dosing upstream of a treatment plant in California was linked to "frequent episodes of effluent quality deterioration."

7.7.3 Liquid Phase Biological Treatment

Methods that seek to modify microbial metabolism, change microbial populations, or otherwise affect the microbes responsible for generating odors in wastewater collection systems are broadly classified as liquid phase biological treatment methods. These are distinct from liquid phase chemical treatment methods, which seek chemical reactions with odor-causing compounds to convert them to nonodorous species.

Two main types of biological treatment methods have been marketed and applied with varying degrees of success. One method, bioaugmentation, seeks to deluge a collection system with bacteria specially bred to oxidize sulfide as it is generated, thereby reversing the natural anaerobic conversion of sulfate to sulfide. The second method, enzymatic treatment, uses substances that either interrupt the bacterial metabolic process that converts sulfate to sulfide or enhance enzymatic activity of microbes that oxidize sulfide.

7.7.4 Bioaugmentation

Four papers in the literature search provided information on the effectiveness of bioaugmentation. Sercombe (1995) studied several liquid phase treatment methods for reducing hydrogen sulfide at six locations within a large sewer network in England. A proprietary brand of bacteria called Bio-Act was applied at one of the treatment works over six months. An initial large dose used to establish the bacteria was followed by lower maintenance doses. The paper does not detail which processes were dosed or wastewater flow rates. However, sulfide was reduced from 4 mg/L to an average concentration of 1 mg/L. Also, H_2S gas was reduced from 500 ppm to less than 100 ppm in the primary-tank desludging chamber.

In another bioaugmentation study, Richman (1997) described a treatment program used in Tampa, Florida, to reduce odors by using a combination of ozonation and a patented in-pipe technology. The city claims to have saved $175,000 in odor control costs. However, the paper does not differentiate between savings due to ozonation and savings due to bioaugmentation.

Nowak et al. (2004) reported on work in Vienna, Austria, collection systems where using bacterial preparations was attempted. The specific products were not identified, but the study indicated that none of the products achieved significant H_2S or odor reduction. Van Durme and Berkenpas (1989) reported on trials of bacterial cultures in Arizona collection systems. They reported no reductions in either dissolved or gaseous sulfide and no noticeable downstream impacts (positive or negative) to the treatment plant.

7.7.5 Enzymatic Treatment

One of the best-known enzymatic treatment methods is anthraquinone, an aromatic ketone reported to block metabolic pathways for sulfate-reducing bacteria. Schneller et al. (2001) reported that anthraquinone binds with a protein that is unique to sulfide-reducing bacteria. Harshman (2003) more specifically reports that anthraquinone interrupts sulfate reduction by interacting with the cytochrome c3 location in the cytoplasmic membrane.

Anthraquinone is also added to some of US Filter's nitrate-based products as an enhancement and distributed under the name Bioxide-AQ. Two field studies testing anthraquinone alone report that it had no effect on dissolved sulfide concentration (Hamilton et al., 1992; Van Durme and Berkenpas, 1989). Wolstenholme et al. (2004) reported on a comparison of Bioxide (calcium nitrate) alone to the combination of Bioxide and anthraquinone (Bioxide-AQ). This work was done on a force main discharge from a naval station receiving waste from docked naval ships. Although dosing duration was not discussed, this paper indicated that when anthraquinone was added, there was no significant difference reported in the odor levels at the force main discharge.

Hamilton et al. (1992) reported that anthraquinone dosed at 10 mg/L inhibited sulfide generation in laboratory tests, but follow-up field dosing of up to 480 mg/L provided no reduction in dissolved or gaseous sulfide on collection systems in southern California. In this field testing the anthraquinone was only dosed for 30 minutes. This may indicate that batch treatment of anthraquinone is not effective.

In one of the few studies in which a biological treatment method was tried alone, Onondaga County Department of Drainage and Sanitation (2001) reported the results of treating a sanitary sewer with an enzyme blocker called Enzybate-E. Enzybate-E was dosed at 20 ppm for approximately 1 month. While some sulfide reduction was observed, an optimal dose could not be determined from the study, and the product was not determined to be cost-effective for controlling sulfide.

OrTec is a product that uses enzymes and nutrients to enhance the ability of bacteria to oxidize odor-causing compounds. In two studies (OrTec, 2002a, 2002b), wastewater was sparged with nitrogen while DO and OrTec concentrations were varied. The studies showed a decrease in ammonia and D/T concentrations due to the OrTec addition. However, these experiments were carried out in small test tubes for over 30 days, and results were not separated from the effect of BOD consumption. Kim (2004) studied the ability of OrTec to inhibit biofilm development on a column of glass beads dosed with wastewater and OrTec. The treated column had a biofilm mass reduction of 50% compared with the control.

Based on this review, the effectiveness of these types of compounds is unclear at best and requires further investigation. Pilot testing would be required in all cases.

7.8 Dosing Summary and Supplier Information

Table 7-2 summarizes dosage information for each liquid phase odor control method. Some chemicals have a wide dosage range. This is representative of the varying results of field- and bench-scale studies. Closer estimates of chemical doses needed for certain applications should be obtained by running a bench-scale test on actual wastewater samples. If possible, bench-scale testing is recommended.

Table 7-2. Dosage Information from Liquid Phase Odor Control Methods.

Chemical	Concentration (% by weight)	Solution Density (lb/gal.)	Dose (lb/lb of Sulfide Removed)	
			Low	High
Pure O_2	95	gas	Not directly comparable to other chemicals	
Cl_2 gas	100	gas	4	15
NaOCl	15	9.7	10	15
H_2O_2	50	10.0	1	4
$KMnO_4$	5	8.7	6	7
$NaMnO_4$	20	9.8	7	8
$FeCl_2$	30	10.0	4	12
$FeSO_4$	6	10.0	5	14
$FeCl_3$	40	10.8	3	7
$Ca(NO_3)_2$	60	12.1	4	15
NaOH	50	11.4	Doses must be compared to other methods on a case-by-case basis as they are not directly related to sulfide concentration alone	
$Mg(OH)_2$	63	13.3		

7.9 Research Agenda Items

Several potential research gaps were identified with relationship to liquid phase treatment. They are highlighted at the end of this chapter. More-detailed research gap ranking and testing protocols are provided in the final chapter of this report.

7.9.1 Life-Cycle Analysis of Collection System Chemical Addition Impacts on Downstream Processes

Liquid phase treatment of wastewater to remove dissolved sulfides usually involves adding significant quantities of chemicals; several classes of chemical are used. Chemical oxidants (for example, hydrogen peroxide and sodium hypochlorite) are used to oxidize sulfide to sulfate or sulfur, iron salts are used to precipitate sulfide to an insoluble solid form, nitrate is added to prevent sulfide from forming, and pH adjustors (for example, magnesium hydroxide) are used to prevent sulfide from entering the gas phase.

Many of the chemicals used in liquid phase treatment may have side-effects on the downstream treatment plant and receiving waters. For example, chlorine used to oxidize sulfide may have a residual disinfectant effect on activated sludge processes, and iron salts can increase sludge production. Nitrate addition may increase the nitrogen load on the plant. Some chemical suppliers claim that their products have beneficial side effects on downstream processes. For example, a supplier of magnesium hydroxide claims that their product improves flocculation in the primary clarifier.

7.9.1.1 Problem Statement

Although effects of chemical addition on downstream processes have been reported anecdotally, little quantitative information on these effects is available. In particular, the life-cycle analysis (LCA) of protecting infrastructure and preventing emissions verses downstream wastewater treatment impacts are unknown. A study is needed to quantify the downstream effects of adding chemicals to the collection system and compare their performance using an LCA framework that calculates technical options based on economic, social, and environmental parameters. LCA is an ISO-standardized framework for the comparison of alternative products or services from an environmental perspective. The fundamental idea behind LCA is much broader than that behind life-cycle costing and consists of assessing all the physical and energetic flows that occur as a consequence of the selection of a particular process (see below).

7.9.1.2 Research Proposal

The objectives of the proposed study are the following: 1) Identify effects on the downstream processes of commonly used chemicals added to collection systems; 2) develop LCA framework; and 3) quantify effects of chemical dosing on wastewater treatment using LCA.

7.9.2 Pure Oxygen Treatment of Gravity Sewers

Pure oxygen injection has been employed to prevent sulfide generation in force mains. Pressurized pipes flowing full allow much higher oxygen concentrations (and fewer injection stations) than can be applied in gravity sewers. Because of the high unit costs for transporting oxygen and high capital costs for onsite generation, oxygen injection into gravity sewers is usually ruled out. However, small-scale pressure swing adsorption (PSA) could enable oxygen injection spaced frequently enough along gravity sewers to prevent anaerobic conditions from developing between injection points.

7.9.2.1 Problem Statement

If oxygen-generating equipment could be made small enough, oxygen injection could save money by replacing chemical addition in gravity sewers. A study is needed to determine the feasibility of developing small PSA units for frequent spacing along gravity sewers.

7.9.2.2 Research Proposal

The objectives of the proposed study are the following: 1) Calculate the range of oxygen generation capacity appropriate for controlling sulfide generation in 0.5-mgd and larger gravity sewers; 2) identify technical and market obstacles preventing PSA manufacturers from developing small-scale PSA systems; and 3) determine the feasibility of introducing a small-scale PSA system to the odor control market.

7.9.3 Magnesium Hydroxide Odor Control Mechanisms Other Than pH Adjustment

Magnesium hydroxide $Mg(OH)_2$ is a common chemical used to prevent sulfide release in sewers. It works primarily as a pH adjuster, raising wastewater pH to between 8.5 and 9. In this pH range, sulfide tends to remain in solution, where it does not cause odor and corrosion problems. Magnesium hydroxide is less soluble, however, than other pH adjusters, for example, caustic soda and lime. As such, it buffers the pH rather than quickly reacting, a favorable characteristic.

Whereas the pH adjustment effects of magnesium hydroxide are generally accepted in engineering practice, other, less-well-understood benefits have been observed and attributed to the chemical. Premier Chemical, a supplier of magnesium hydroxide, claims that sulfide complexes with Mg^{2+} ions form a suspension that keeps its form during treatment processes. This action is purportedly similar to sulfur precipitation by iron salts and results in sulfide removal as well as pH adjustment. Premier Chemical also claims that Mg^{2+} catalyzes the oxidation of dissolved sulfides. The rate of sulfide oxidation is believed to increase approximately two to three times in the presence of magnesium ion.

7.9.3.1 Problem Statement

Although sulfide removal by magnesium hydroxide has been observed anecdotally, independent research has not been carried out to verify the claims of chemical suppliers. Therefore, an independent study is needed to establish the mechanisms responsible for sulfide removal by magnesium hydroxide and to quantify dosages that can be used to design chemical treatment systems.

7.9.3.2 Research Proposal

The objectives of the proposed study are the following: 1) Experimentally determine chemical mechanisms responsible for sulfide removal by magnesium hydroxide; 2) verify or debunk existing theories about sulfide removal mechanisms; and 3) quantify dosage rates for sulfide removal by magnesium hydroxide.

7.9.4 Liquid Phase Biological Treatment

Methods that seek to modify microbial metabolism, change microbial populations, or otherwise affect the microbes responsible for generating odors in wastewater collection systems are broadly classified as liquid phase biological treatment methods. These methods are distinct from liquid phase chemical treatment methods, which seek to chemically react with odor causing compounds and convert them to nonodorous species.

Three main types of biological treatment methods have been marketed and applied with varying degrees of success. One method, bioaugmentation, seeks to deluge a collection system with bacteria specially bred to oxidize sulfide as it is generated, thereby reversing the natural anaerobic conversion of sulfate to sulfide. Another method, enzyme blocking, uses substances to interrupt the bacterial metabolic process that converts sulfate to sulfide. The third method, nitrate addition, seeks to supply sulfate-reducing bacteria with an alternate electron acceptor, whereby bacteria convert nitrate to N_2 gas instead of sulfate to sulfide.

7.9.4.1 Problem Statement

Whereas the effectiveness of nitrate addition is well documented both in practice and in the technical literature, the effectiveness of the other two biological treatment methods has not been sufficiently verified or documented. In some reports, the effectiveness of biological methods (other than nitrate) was inconclusive (Onondaga County Department of Drainage and Sanitation, 2001). In others, biological treatment methods were tested concurrently with nitrate or chemical methods, and the contribution of bioaugmentation or enzyme blocking could not be determined (Arthur and Anker, 2000; Kim et al., 2005). In still other studies, anecdotal reports of the effectiveness of biological treatment methods are provided without showing data or system characteristics (Sercombe, 1995).

Owing to the prolific patenting and marketing of bioaugmentation and enzyme-blocking products, there is a need for systematic research verifying the effectiveness, or lack thereof, of these biological methods. Needed is an independent study that tests leading biological methods under realistic conditions, without the influence of other methods. All relevant parameters should be reported and the duration of the study should allow sufficient time to arrive at conclusive results.

7.9.4.2 Research Proposal

Objective The objectives of the proposed study are the following: 1) Identify the leading biological treatment methods (excluding nitrate addition, which has already been sufficiently tested); 2) conclusively determine the level of effectiveness of each leading product, 3) and determine the applications and doses for products determined to be effective.

Program The program of research would consist of the following three phases: study design, pilot-scale testing, and results documentation.

CHAPTER 8.0

GAS PHASE ODOR TREATMENT

8.1 Overview of Literature Reviewed

Fifty-one papers, articles, reports, and theses, covering a wide range of gas phase odor treatment technologies, were reviewed. The breakdown of papers by topic was as follows:

- Biofilters (13)
- Biological scrubbers (11)
- Membrane bioreactors (1)
- Activated-sludge diffusion (3)
- Chemical scrubbers (4)
- Activated carbon (5)
- Ionization (4)
- Cold plasma reactors (3)
- Photocatalytic reactors (1)
- Overviews (6)

Figure 8-1 shows a distribution of topic areas. The general topic of biological treatment of odors represented about 53% of the papers reviewed; most of the papers were published within the past five years.

Of the 51 papers, 29 were found in technical journals or conference proceedings, and eight in "soft journals" such as equipment trade magazines; three were theses, and one was a technical report for a municipal client (that is, gray literature). Figure 8-2 shows the distribution of papers by type of literature.

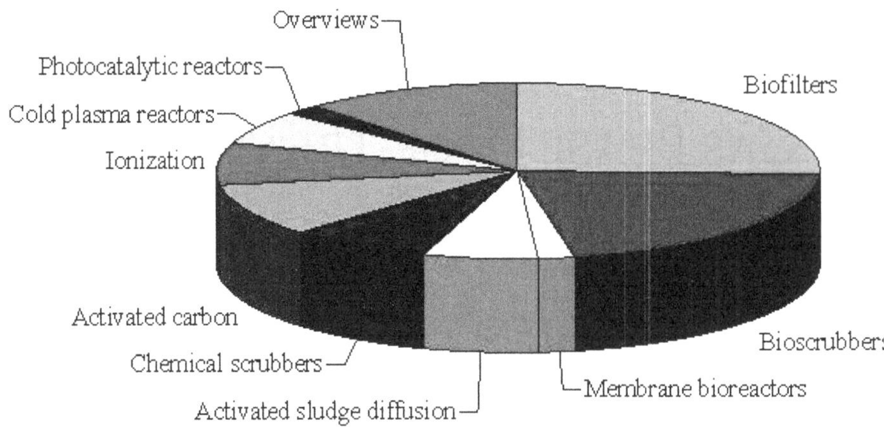

Figure 8-1. Distribution of Topic Areas.

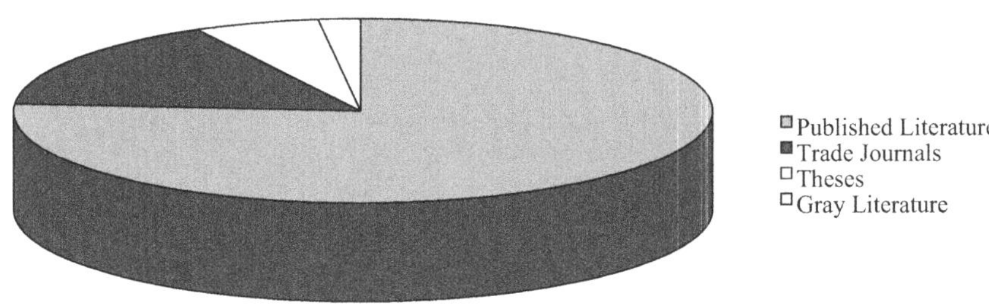

Figure 8-2. Literature Distribution by Source.

8.2 Background

In collection systems, controlling odors sometimes requires "capture-and-treat" approaches, whereby a fan is used to pull odorous air from a structure and convey it to an odor treatment device. This is as opposed to the direct addition of chemicals to oxidize, precipitate, or prevent the release of odors and H_2S. The capture-and-treat approach is often used at sensitive locations where the free flow of air down a sewer is blocked by either 1) the discharge of the sewer into a pump station wet well, 2) an inverted siphon without an air jumper—or with an inadequately sized one—or 3) inadequate sewer headspace. Because of the drag exerted by the flowing sewage on the headspace air, the air travels down the pipe until it reaches an obstruction. The lack of an outlet for this air causes pressurization of the headspace, forcing the odorous gases out vents, cracks, seams, etc. This phenomenon also occurs at a significant grade change, such as when sewage travels down a steep sewer that "flattens out" at the bottom of the hill. Air pressure is negative upstream of the slope change, causing air to be drawn into the sewer, but is positive after the slope change, forcing odors out of manholes, etc. In collection systems, the most common application of odor control technology is for pump station wet wells. More-extensive odor control may be necessary at pump stations that remove and process grit and screenings. Other than pump stations, capture-and-treat systems are sometimes used to control odor emissions from siphon head structures, junction boxes, and drop structures, where headspace pressurization and odor release occur.

Ventilation rate is an important parameter in sizing an odor control system. Unfortunately, air movement patterns in sewers and sewer structures are not well understood and are rarely measured, making difficult the selection of the proper airflow to achieve a negative pressure. In some cases, standard air exchange criteria such as "12 air changes per hour" are applied to wet wells that are entered by workers. However, such rules of thumb often have little basis in achieving the main objective in sewer ventilation: preventing the escape of odors by maintaining a slight negative pressure. For covered processes at wastewater treatment facilities, some practitioners have proposed ventilation rates in proportion to the covered surface area, with the range dependent on the "tightness" of the cover and the degree of infiltration through cracks and openings. Guidelines range from 0.2 to 0.5 $m^3/min/m^2$ of covered surface area (0.5 to 1.5 cfm/ft^2).

There are many technologies used to treat odorous air. Below is a review of those known to be successful for odor control, along with some that are in the research and development stage. For each technology, a brief description is provided, followed by a discussion of application, performance, and development status.

8.3 Chemical Scrubbers

8.3.1 Description

Packed-tower, wet-chemical scrubbers have long been the workhorse of the odor control industry. Chemical scrubbing is a fully proven technology that performs effectively and predictably. Chemical scrubbers consist of corrosion-resistant vessels in which odorous air contacts a chemical solution to absorb or oxidize odorous constituents. The most common configuration of a chemical scrubber consists of a vertical tower filled with plastic packing in which contaminated air flowing upward contacts a chemical solution flowing downward (Figure 8-3). In wastewater treatment applications, the chemical solution typically consists of sodium hypochlorite (bleach) or sodium hydroxide (caustic soda) or both. In the majority of cases, chemicals are metered into a scrubbant recycle line to maintain pH and oxidation-reduction potential (ORP) set points. Proper levels of chemicals control the absorption and oxidation of H_2S and other reduced-sulfur compounds. Chemical scrubbers have a small footprint and can be constructed to perform in multiple stages, depending on the types and levels of contaminants to be removed.

8.3.2 Application

Chemical scrubbers are widely used for treating odors from wastewater treatment facilities. Although chemical scrubbers can achieve high odor removal efficiency for a wide range of odorant loadings, the systems require a significant level of operation and maintenance. In collection systems, this has generally limited their application to larger pump stations that are routinely monitored by operations staff. Further, the systems require the storage and metering of chemicals such as sodium hypochlorite (bleach) and sodium hydroxide (caustic soda) that are often undesirable in residential neighborhoods. Although "package" scrubber systems are available for small pump stations, most municipalities choose systems that require less operator attention, such as biofilters and activated carbon adsorbers, because facilities in the collection system may be checked only intermittently by operations staff.

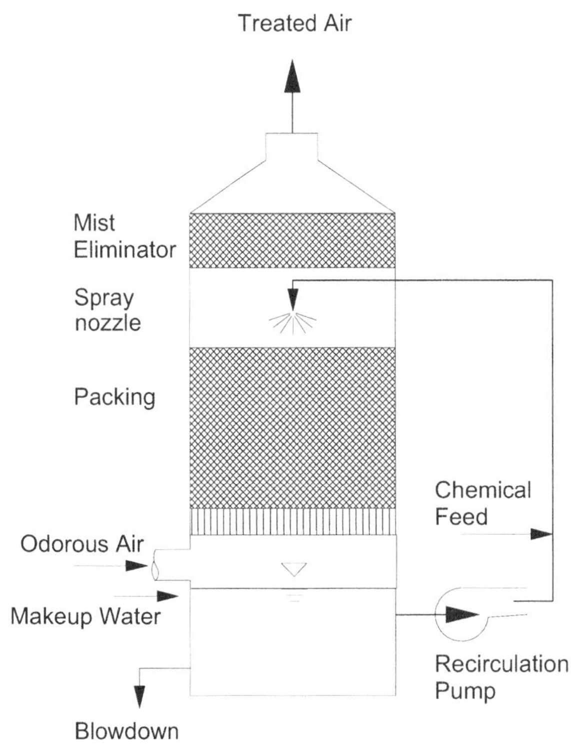

Figure 8-3. Packed Tower Chemical Scrubber.

8.3.3 Performance

Chemical scrubbers using sodium hypochlorite and sodium hydroxide remove H_2S and many odorous sulfur compounds very well; those using sulfuric acid remove ammonia and trimethyl amine very well. Chemical scrubbers' performance is predictable, and the makeup of the chemical solution can be modified easily by changing the set points. Odor removal efficiency typically exceeds 90%, and H_2S removal can be 99.9%.

8.3.4 State of the Art

Of the four papers on chemical scrubbers reviewed, one dealt with using hydrogen peroxide instead of sodium hypochlorite as a scrubbant chemical, one addressed operational issues with scrubbers, one investigated the formation of chlorinated organics, and one was a vendor-prepared case study. Two of these papers also dealt with converting a chemical scrubber to a biotrickling filter. The processes that occur in chemical scrubbers are well researched and documented. With the exception of new methods to control chemical feed rates (for example, using outlet H_2S concentration rather than pH/ORP) or novel "package" configurations of multistage systems, there have been relatively few developments in the wet-scrubber industry in the last 20 years. Traditional pH/ORP control of chemical feed rates is still the standard control system, but fundamental problems with this method remain. These include the lag time between the chemical dosing point and the location where pH/ORP is monitored and the difficulty in maintaining a stable operation under variable loading conditions.

8.4 Activated-Carbon Adsorbers

8.4.1 Description

Activated-carbon adsorbers, like chemical scrubbers, are a traditional workhorse of the odor control industry. Activated carbon adsorbers are simple systems consisting of a corrosion-resistant vessel containing an activated carbon medium through which odorous air is passed to remove odorous contaminants by adsorption. Systems configurations can be "single-bed" or "dual-bed," with each bed of carbon being about 3 ft thick. In dual-bed systems, the inlet airflow is split, and air passes through the two beds in parallel (one air stream upward, one downward) and is exhausted through stacks at the top. Figure 8-4 shows a typical dual-bed activated-carbon adsorber. For wastewater treatment applications, the activated-carbon medium is often a "specialty" carbon designed for H_2S and sewage odors (Section 8.4.4). For relatively dilute odors or for polishing the exhaust from other odor control processes, "virgin" coconut-shell or coal-based carbon can be used.

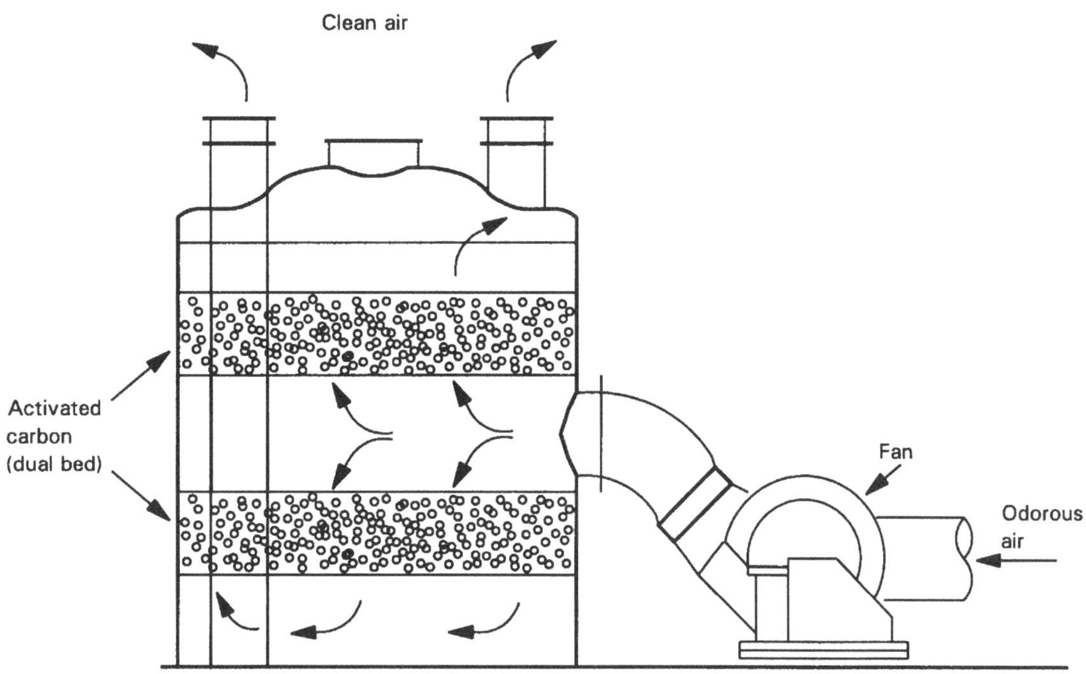

Figure 8-4. Dual-Bed Activated-Carbon Adsorber.

8.4.2 Application

The simplicity of an activated carbon system favors its use at pump stations or other unstaffed sites in the collection system—electric power is all that is required for utilities. Carbon is generally effective for a wide range of odorants. There are many vendors of activated-carbon systems, and the performance and longevity of the medium is relatively predictable. New developments in carbon media have extended the range of H_2S loadings for which carbon is appropriate. Activated carbon continues to be widely used for odor control applications at wastewater collection and treatment facilities.

8.4.3 Performance

Carbon performs very well in wastewater odor control applications; it is generally effective for a wide range of odorants, and over the past 20 years, significant improvements have been made to the medium so that it more effectively removes hydrogen sulfide, a main component of sewage odors (especially in collection systems). The disadvantage of carbon is that high-odorant loadings may cause breakthrough and rapid media exhaustion, thus making frequent media regeneration or replacement necessary.

8.4.4 State of the Art

In the past 10 years, there have been significant advances in activated carbon designed for H_2S removal. The traditional approach had been to impregnate the medium with sodium hydroxide or potassium hydroxide to improve H_2S removal (via absorption/reaction). Unfortunately, the medium was subject to spontaneous combustion and plugging, and it required disposal as a hazardous waste owing to its high pH. The first major innovation was a "catalytic" carbon, developed by Calgon, in which the H_2S is oxidized to sulfate on the surface of the carbon. Its ability to remove H_2S can be restored by soaking or washing it in water to remove the sulfate. More recently, single-use "high-H_2S-capacity" carbon has been developed by US Filter-Westates and others. While not regenerable, its H_2S removal capacity is up to six times that of virgin (untreated) carbon. Table 8-1, with data from Westates, shows the H_2S capacity of various carbons:

Table 8-1. H_2S-Removal Capacity of Various Carbon Media.

	H_2S Removed	
Carbon	Grams per Cubic Centimeter	Percent by Weight
Coconut-shell	0.05	10
Caustic-impregnated	0.14	25
Water-wash catalytic	0.09	16
Single-use, high-capacity	0.30	62

Of the sources on using carbon for odor control that were reviewed, one (a thesis) researched the use of unmodified carbon on odor and H_2S removal, one investigated using activated carbon cloth for odor removal, one (a thesis) researched the removal of carbonyl sulfide, and one evaluated the mechanism of removing biological H_2S in unmodified activated carbon.

New activated-carbon media designed for H_2S and sewage odors has advanced the state of the art for odor treatment via carbon adsorption and thus expanded the range of applications for which carbon treatment is a viable and cost-effective alternative.

8.5 Biofilters

8.5.1 Description

Biofiltration is a technology developed mostly in Europe that has been used widely in the U.S. for the past 15 years. Biofilters are now a viable and accepted alternative to chemical scrubbers and activated-carbon systems. There is a variety of engineered, "in-ground" biofilters as well as proprietary, "in-vessel" biofilters in use at wastewater treatment plants and pump stations throughout North America.

The traditional biofilter consists of an open bed of organic material through which odorous air is passed. The medium is often a blend of wood chips, compost, and bark nuggets designed to maintain porosity while allowing good contact between the odorants and the biologically active bed. Figure 8-5 is a schematic of a traditional biofilter. Soil is also used as a natural medium. One vendor has developed a nondegradable biofilter medium consisting of a porous substrate coated with nutrients, buffers, and adsorbents.

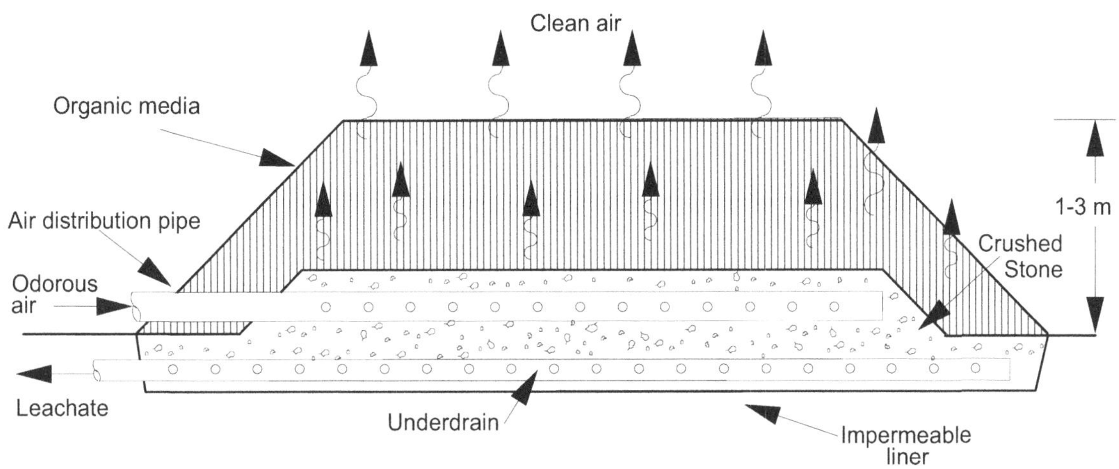

Figure 8-5. In-Ground Organic-Media Biofilter.

Good distribution of air through the media and maintenance of the proper moisture content are essential. Moisture can be applied by prehumidification, sprinklers, and soaker hoses. Vendor-supplied "package" biofilter systems have addressed these issues in the design of the vessels and humidification systems. There are many different configurations of proprietary in-vessel biofilters.

8.5.2 Application

Biofilters are frequently used for odor control at pump stations. Proprietary package biofilter systems may even enjoy greater application than traditional in-ground biofilters owing to their smaller footprint. The need for potable water can limit the application of biofilters for junction boxes, inverted siphon headchambers, etc., where potable water is not available. For the past 15 years, biofilters have been applied to a wide variety of odor control applications at wastewater treatment plants and, to a lesser degree, at wastewater collection systems.

8.5.3 Performance

The performance of biofilters is well documented. Odor removal performance is nearly equivalent to that of other traditional control technologies, such as wet scrubbers and activated carbon. Properly designed biofilters can do an excellent job removing both H_2S and reduced-sulfur compounds, although systems often produce a slight residual "earthy" odor. While there is little information in the literature on specific applications of biofilters in wastewater collection systems, there are substantial performance data available on the use of biofilters to treat wastewater treatment plant odors.

8.5.4 State of the Art

Of the 51 papers reviewed, 15 were about biofilters or included a discussion of biofilters. Quigley et al. (2004) provided an excellent summary of the performances of different biofilter media. Shareefdeen et al. (2004) addressed H_2S removal by biofilters in cold climates and provided a case study on removing reduced-sulfur compounds. Three papers described development of biofilter models. Converse et al. (2003) researched volatile organic compound (VOC) removal in biofilters.

There is a significant body of information throughout the world on the use of biofilters for treatment of odorous air. Although there continue to be failures of biofilters due to misapplications, design errors, and poor operation and maintenance, knowledge of biofilters is comprehensive and continues to improve. Engineered in-ground biofilters using organic media are usually very conservatively designed.

Most biofilters are designed with organic media—blends of compost, wood chips, bark, peat, and other materials—which are subject to natural biological degradation and compaction. This is a significant limitation to organic-media biofilters. One vendor has developed a nondegradable biofilter medium that carries a 10-year warrantee, and there has been research published over the years on using perlite, rock wool, sand, and other nondegradable materials, but the vast majority of biofilters in the United States use a blend of wood-based organic media. This is one area where further research and development may be warranted to overcome the limitations of conventional organic media.

8.6 Biological Scrubbers

8.6.1 Description

Biological scrubbers—also referred to as biotrickling filters—consist typically of vertically oriented vessels containing inert packing or media that support biological growth. The vessels are equipped with spray nozzles to continuously or intermittently apply water and nutrients and an underdrain system to collect and in some cases recirculate the leachate. A typical biological scrubber is shown in Figure 8-6. Most are designed for countercurrent operation, with the air flowing vertically upward while the water trickles downward, though one vendor uses a cocurrent design. Another version of this concept uses a combination of attached biological growth on the media with a small, suspended-growth reactor in the recycle stream. The growth on the fixed media allows absorption and oxidation of odorants as the air passes through the media. The suspended growth reactor is used for the oxidation of odorants after transfer into the scrubbant liquid. Most biological scrubber systems are relatively simple, with few moving parts. The advantage of the biological scrubber over a biofilter is that the vertical orientation of the vessel offers a potentially smaller footprint than most biofilters.

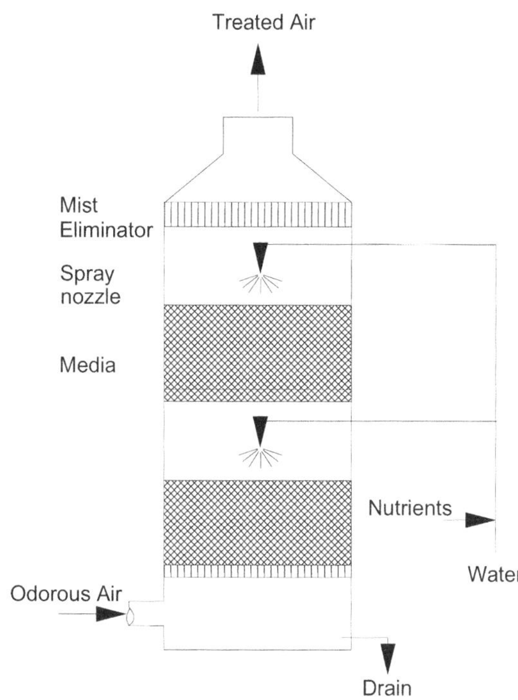

a. BIO-TRICKLING FILTER

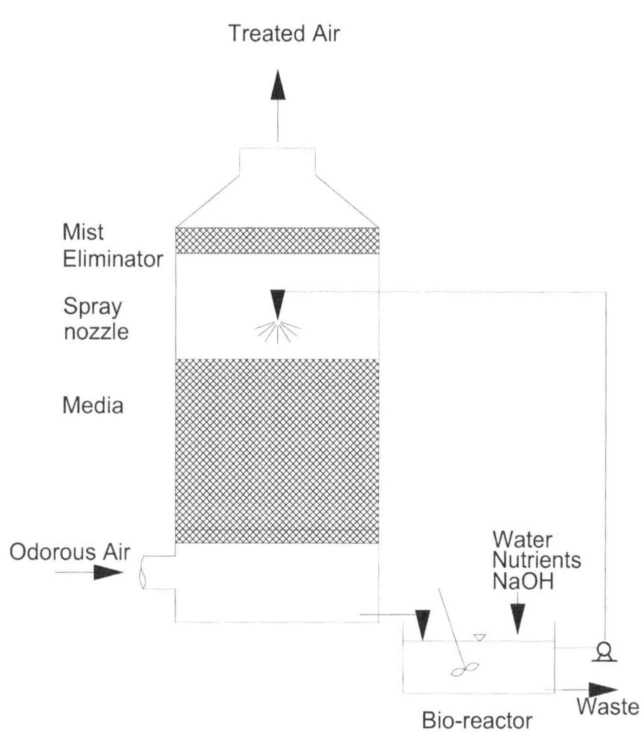

Figure 8-6. Biological Scrubber Systems: Biotrickling Filter (top) and Bioscrubber.

8.6.2 Application

Although most biological scrubbers in the United States have been installed at wastewater treatment plants, the technology is suitable for use in collection systems, particularly at pump stations. Biological scrubbers are low maintenance and require little operator attention. As described in the next subsection, the systems produce a musty or earthy odor that depending on the proximity of receptors may require additional polishing.

8.6.3 Performance

Biological scrubbers perform extremely well on hydrogen sulfide, with an H_2S-removal efficiency over 99% in some cases. Easter et al. (2005) reported an average of 98% H_2S removal from eight biotrickling filters. However, average *odor* removal was only 71%. Increasing the empty-bed residence time (EBRT) increases the efficiency of removing odor and H_2S. Higher EBRTs are necessary to treat significant levels of non-H_2S reduced-sulfur compounds. Biotrickling filters have been very successful at pump stations where H_2S is the dominant odorant.

8.6.4 State of the Art

Interest in biological scrubbers as an alternative to biofilters or conventional odor control technologies has increased dramatically in the last five years. Of the 51 papers reviewed, 13 addressed biological scrubbers (some papers dealt with both biofilters and biological scrubbers). The availability of small-footprint biological odor control systems with nondegradable media is very attractive to users who seek a reliable, low-operations and -maintenance system that does not use hazardous chemicals and does not require frequent media replacement.

Of the papers on biological scrubbers, several were overviews of biological odor control alternatives, three reported on the conversion of chemical scrubbers to biological scrubbers at the Orange County (Calif.) Sanitation District, and three were laboratory investigations on removing specific contaminants, such as H_2S and ethanol; there was one paper each on 1) pilot testing for odor and VOC removal, 2) a full-scale application in Denmark, and 3) the use of "biological activated carbon" as a medium for a biological scrubber.

Although municipalities have converted existing vessels into biological scrubbers or have designed their own biological scrubbers, most rely on vendor-supplied systems. There are at least six vendors of biological scrubbers; three use random-dump polyurethane foam cubes as the medium, one uses woven-plastic "cassettes," one uses foam media custom fit to the vessel dimensions, and one uses lava rock media.

Many of the biological scrubber systems currently operating in the United States are in warm climates, where H_2S is the predominant odorant, and where warm temperatures promote high biological activity and good reaction kinetics. For these applications, H_2S removal has generally been excellent. However, there is a lack of information and good performance data on how these systems function in colder climates, where H_2S may be in relatively low concentrations, or where the predominant odorants may be non-H_2S sulfur compounds.

8.7 Ionization Systems

8.7.1 Description

Air ionization or oxygen ionization is a process by which positive and negative ions are generated with a reactive-plasma generator. The technology has been used in Europe for some 20 years. According to vendor literature, the ions activate clusters of oxygen molecules that react with contaminants in the air. One vendor has postulated that the activated oxygen clusters react with H_2S to form sulfate. Normally, the ion generator is applied to filtered, ambient supply air that is then introduced into an odorous room or process to allow contact between the ions and the odorants. Claimed advantages include low power requirements, small footprint, and low capital and operating costs. Figure 8-7 shows a diagram of the air ionization process.

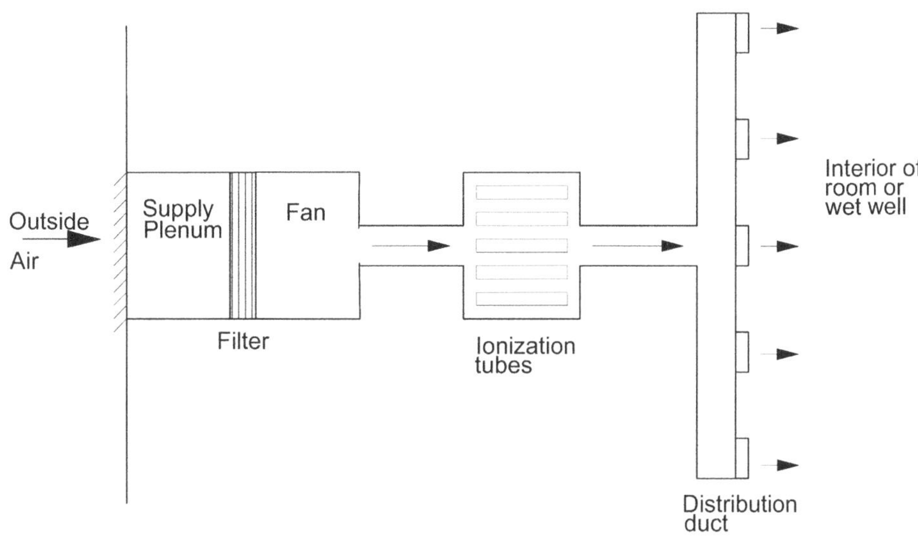

Figure 8-7. Air-Ionization Process.

8.7.2 Application

Air ionization technology is potentially applicable to both collection system and wastewater treatment plant odors. In theory, the relatively small ion generators could be installed in pump stations as a means of treating wet well odors. Most of the existing applications are for treating odors from sludge-handling operations.

8.7.3 Performance

The performance of air ionization applications at wastewater treatment facilities is poorly documented. Of the three papers reviewed in the published literature as part of this project, two discussed a pilot study in Germany whereby the air ionization process was used to pretreat VOCs in odorous air prior to biofiltration; a 25-35% reduction in VOCs was achieved. A third paper was a presentation of case studies by a U.S. vendor. A field "scentometer" was used to measure "before" and "after" odor levels. H_2S measurements showed virtually complete H_2S removal at one site. The validity of the data is unknown.

The most comprehensive and scientific testing of air ionization was conducted by Webster Environmental Associates, Inc. (2003), for the Orange County (Calif.) Sanitation District. Two different vendor-supplied systems were tested. When no supplemental H_2S was added to the room air to be treated, an approximately 50% reduction in odor and H_2S was achieved. This test was done when H_2S concentrations ranged from 0.1 to 0.2 ppm. When bottled H_2S was introduced to achieve a baseline H_2S of 3 to 5 ppm, there was no consistent reduction in odor or H_2S associated with the introduction of ionized air.

8.7.4 State of the Art

Although ionization systems have been used for odor control in Europe for 20 years, there are few scientific data in the published literature to document their effectiveness or the mechanism by which odor reduction occurs. There are relatively few systems in the United States and few data to support the vendors' claims, even though operations staff subjectively report improvements in air quality where the systems are in use.

8.8 Other Biological Systems

8.8.1 Description

Other biological odor treatment systems include diffusion into activated-sludge basins and membrane bioreactors. Activated-sludge diffusion is a process whereby odorous air is introduced into existing blowers and diffused into aeration tanks for treatment. A schematic is shown in Figure 8-8.

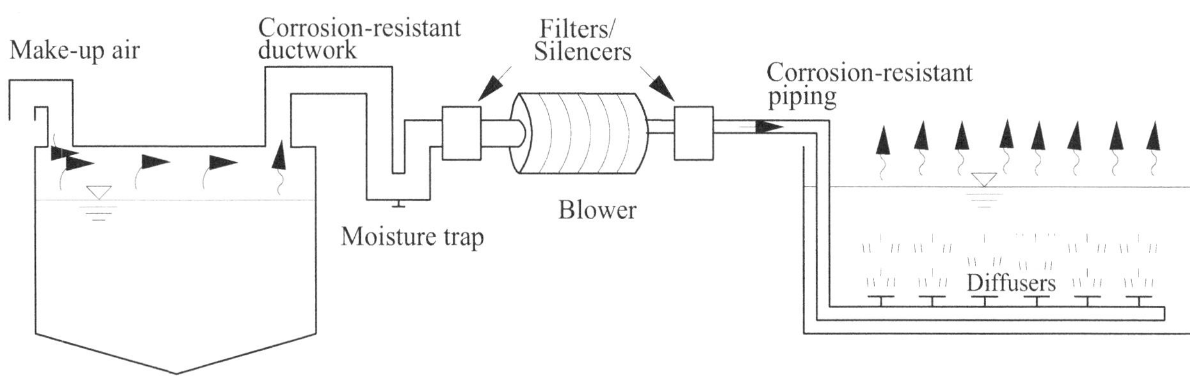

Figure 8-8. Activated-Sludge Scrubbing System.

Membrane bioreactors (MBRs) are a relatively new development in the field of wastewater treatment. Although a number of full-scale applications exist for the treatment of wastewater, using MBRs for control of odorous air emissions is still in the research stage. Figure 8-9 illustrates the concept of an MBR for treatment of odorous air.

8.8.2 Application

Since activated-sludge diffusion relies on the availability of an existing activated-sludge process, it has little application for collection systems, other than perhaps the treatment of air from an influent sewer or influent pump station located at the plant.

MBRs have a potential for use in a wastewater collection system or treatment plant. Given the limited stage of development of this technology for odor control, it is impossible to assess the factors that may affect its application.

8.8.3 Performance

The performance of activated-sludge diffusion as a method of odor treatment is excellent and well-documented. Three of the papers reviewed addressed performance of activated-sludge diffusion for odor control.

Use of membrane bioreactors for odorous air treatment has been researched at laboratory scale. Most studies have been on single contaminants.

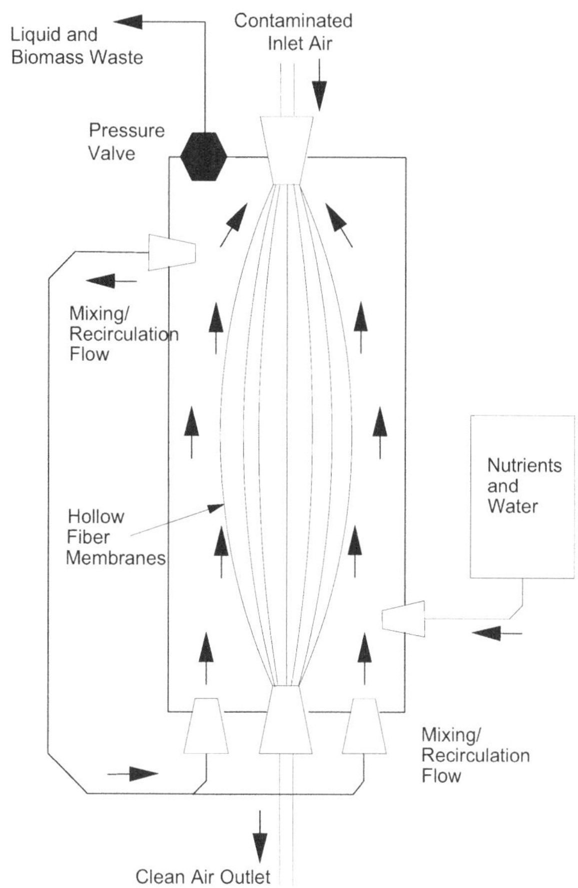

Figure 8-9. Schematic of Hollow Fiber Membrane Reactor.

8.8.4 State of the Art

Activated-sludge diffusion is a well-documented odor treatment technique that has limited application to wastewater collection systems owing to its dependence on an existing activated-sludge basin being close to the odor source. Research to better define process mechanisms and limitations is ongoing.

Using membrane bioreactors to treat odorous air emissions is in the developmental stage, with only laboratory data available on its effectiveness. Only one paper on the subject was reviewed as part of this project.

8.9 Hydroxyl Ion Fog

8.9.1 Description

The hydroxyl ion fog process is a proprietary system that is used for wastewater odor control applications. The system combines onsite-generated ozone, water, and compressed air, which produce an atomized fog of ozone and hydroxyl radicals. According to the vendor, this oxidizing fog has a rapid chemical reaction with H_2S and other odorous compounds. The mechanism for removing hydrogen sulfide and other odorants is not well understood. The fog is blown into a wet well or other contained structure to react with the odorous compounds, eliminating the need for a separate vessel. The system has also been applied to ductwork upstream of a wet scrubber. A diagram of the process is shown in Figure 8-10.

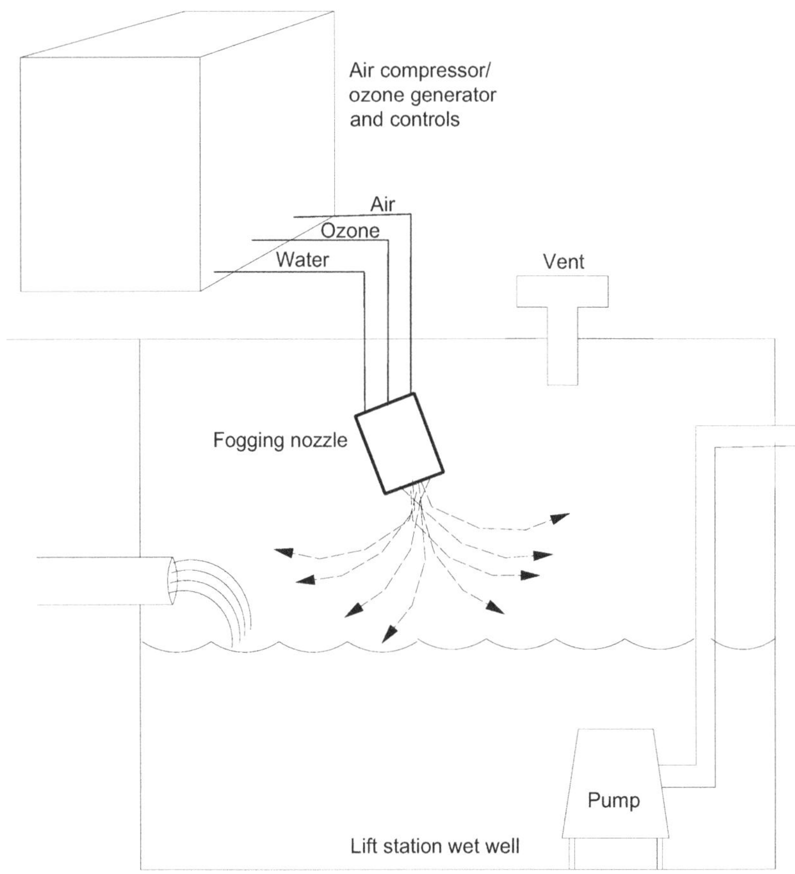

Figure 8-10. Hydroxyl Ion Fog Process.

8.9.2 Application

The hydroxyl ion fog system requires water and electricity. There are units operating on pump station wet wells as well as wastewater treatment processes. The Hampton Roads (Va.) Sanitation District has several units operating at its facilities.

8.9.3 Performance

No papers were reviewed on this technology as part of the WERF project. Two papers on the subject were presented at the 2002 WEF Odor Specialty Conference. Good performance data are lacking. Limited H_2S data were presented, but no data on odor reduction or the removal of non-H_2S reduced-sulfur compounds were included.

8.9.4 State of the Art

The hydroxyl ion fog process has been in use for approximately eight years. Unfortunately, there are very few published data on the effectiveness of the process. Without such data, it is difficult to assess the technology or its application for collection system odor control.

8.10 Cold Plasma and Photocatalytic Reactors

8.10.1 Description

Corona reactors, also called nonthermal plasma or cold plasma reactors, are being studied as a method to treat odorous air. The process involves passing the contaminated air through a reactor receiving a high-voltage electrical pulse discharge. The process does not appear to be in commercial use for odor control applications.

Another system, called a photocatalytic reactor, has also been investigated on a laboratory scale as an odor control method. Here, odorous air is circulated through a reactor containing a titanium dioxide-coated photocatalyst deposited on glass supports. The reactor contains a medium-pressure mercury lamp that activates the catalyst, which generates hydroxyl radicals that react with the odorants.

8.10.2 Application

Because of the limited stage of development, application of the technology to full-scale treatment of municipal wastewater odors is difficult to assess. The process could conceivably be used for odor treatment at both wastewater collection and treatment plants.

8.10.3 Performance

Three papers were reviewed on laboratory testing of cold plasma reactors to treat dimethyl sulfide; methanethiol and hydrogen sulfide; and ammonia, ethanethiol, and trimethylamine. This research was conducted in Taiwan. The work evaluated the effect of inlet concentrations, applied power, and single versus mixed malodorants on performance. The laboratory-scale systems were able to achieve high destruction efficiencies for reduced sulfur compounds and nitrogen-based odorants.

The photocatalytic reactor studies in France investigated the rate of decomposition of a single odorant, 2-butanone. The presence of water vapor increased the production of hydroxyl radicals that enhanced degradation rates.

8.10.4 State of the Art

Cold plasma reactors and photocatalytic reactors are in the early stages of development, and it may be years before the technology is commercialized for municipal odor control applications. There are limited performance data available, and the economics of the processes are unknown at this time.

8.11 Summary

Table 8-2 summarizes gas phase odor treatment technologies, showing the status of development, the availability of performance data, and the applicability to treatment of odors from wastewater collection systems.

Table 8-2. Summary of Gas Phase Odor Treatment Technologies.

Technology	Development Status	Availability of Performance Data	Applicability to Collection Systems
Chemical scrubbers	Fully developed	High	Limited
Activated carbon	Fully developed	High	High
Biofilters	Fully developed	High	High
Biological scrubbers	Evolving	Medium	High
Ionization	Evolving	Low	Limited
Activated sludge diffusion	Fully developed	Medium	Not applicable
Membrane bioreactors	Research	Low	Potential
Hydroxyl ion fog	Developed	Low	Medium
Cold plasma reactors	Research	Low	Potential
Photocatalytic reactors	Research	Low	Potential

8.12 Research Agenda Items

8.12.1 Identifying Compounds in Sewer Gas That Lead to Odor Complaints

Hydrogen sulfide is widely regarded as the most significant odorous compound present in sewer gas and the primary cause of odor complaints associated with wastewater collection systems. This proposition, however, has not been rigorously tested. Control of hydrogen sulfide in many circumstances may not necessarily lead to significant reduction of an odor's intensity or its objectionable nature. Anaerobic decomposition produces a spectrum of intermediate metabolites and end products that have low odor thresholds. These products include short-chain fatty acids, ammonia, organic amines, indole-related compounds, phenolics, and organosulfur compounds. The importance of these compounds in generating odor complaints, relative to hydrogen sulfide, has never been systematically evaluated. Collection system operators have a need to know whether an odor problem they are attempting to correct is likely to be ameliorated by reducing or removing hydrogen sulfide from the sewer atmosphere or whether the odor control strategy must be targeted against other compounds.

8.12.2 Effectiveness of Treatment Methods for Odor Compounds Other Than H_2S

Measurement and odor treatment techniques are usually designed for hydrogen sulfide. However, numerous less-well-understood compounds contribute to odor and warrant attention. Since treatment technologies are geared to hydrogen sulfide, other compounds are less effectively treated and can represent the bulk of odors remaining after treatment.

Neglecting the spectrum of odor-causing compounds has led, in some cases, to the failure of odor control systems even while hydrogen sulfide is effectively treated. Likewise, success of many odor control systems may depend on the absence of other compounds in significant concentrations. A study is needed to evaluate how effectively odor compounds other than hydrogen sulfide are controlled with existing treatment systems.

8.12.3 Effectiveness of Biotechnology in Cold Climates

The effectiveness of biotechnology is somewhat dependent upon a normal range of operating temperatures. Below that range, odor compound removal efficiency can be quite sensitive to temperature. Shareefdeen et al. (2004) studied biofilter removal of hydrogen sulfide at low temperatures in a laboratory and found that removal began to decrease precipitously below 5°C. Moreover, cold weather can cause inlet air to lose humidity, due to condensation, before entering the media. This hampers the effectiveness of biofilters.

Because laboratory experiments indicate that cold temperatures hamper biotechnology effectiveness, full-scale biotechnology systems should be evaluated for their effectiveness in cold weather environments. The study should identify locations where odor control functions properly during warm weather but fails during the winter. These cases should be compared to similar cases where the systems do not fail. Also, the effectiveness of modifications such as insulation and steam addition should be evaluated.

CHAPTER 9.0

CORROSION MECHANISMS AND PROTECTION

9.1 Literature Search Background Summary

Sixty-six papers providing information on corrosion mechanisms and protection in collection systems were located through the literature search. Of these, 26 addressed the mechanism of corrosion, and 41 the protection of pipelines or concrete in general. Some papers compared various corrosion control approaches.

Figure 9-1 shows the distribution of papers by source.

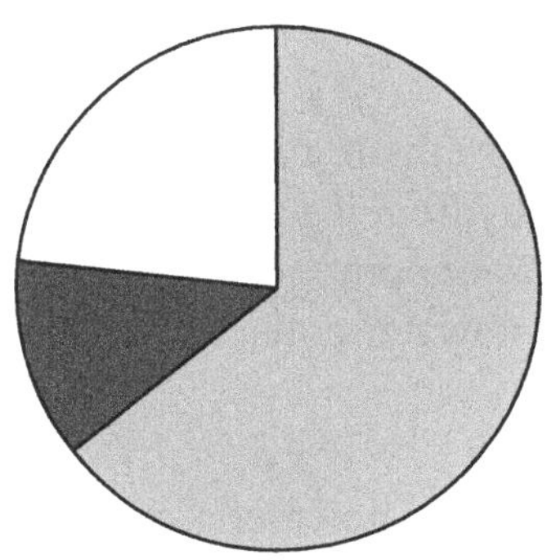

Figure 9-1. Corrosion Mechanisms and Protection Literature by Source.

Papers covered experiences as far back as 1990, with many covering both mechanisms and protection. A good, overall, "plain English" description of corrosion mechanisms and of several coating and lining methods was found in an article by Tator (2003). Additional valuable information is available from well-established reference books and texts, such as the following:

- U.S. EPA (1985) Design Manual 625: *Odor and Corrosion Control in Sanitary Sewerage Systems and Treatment Plants*
- *Hydrogen Sulphide Control Manual* (Melbourne Water, 1989)
- WEF (1995) Manual of Practice 22: Odor Control in Wastewater Treatment Plants
- Odor and VOC Control Handbook (Rafson, 1998)
- WEF (2004) Manual of Practice 25: *Control of Odors and Emissions from Wastewater Treatment Plants*

9.2 Introduction and Background

There has been corrosion of sewer pipelines for as long as sewer pipelines have been in service, and pipeline corrosion was reported as long ago as the end of the 19th century (Olmstead and Hamlin, 1900). But pipeline corrosion has increased noticeably in recent decades because of trends and policies applicable to wastewater collection systems. Two such examples are pretreatment regulations instituted in 1980 and infiltration/inflow (I/I) regulations and policies. Pretreatment has decreased the concentration of metals in collection systems, metals that would otherwise have reacted with and bound up corrosion-causing sulfide compounds, and I/I, which dilutes wastewater and thus decreases odor and corrosion, has been reduced in collection systems.

Corrosion mechanisms have been discovered and studied relatively recently, and measures to control such corrosion have been developed in parallel. Of the many methods currently available to repair, rehabilitate, and replace wastewater collection and conveyance systems are these three: adding coatings, most frequently on the crown of the pipe; lining the pipe wall; and installing pipes made of corrosion-resistant material. The first two measures are generally used for existing pipeline repair or rehabilitation, and the latter for replacement or new pipelines. Because the suitability of a given technology for a specific project is always a function of many varying factors, each method has its appropriate uses and its limitations. Categories of control discussed in other chapters of this report include chemical addition and ventilation.

9.3 Corrosion Mechanisms

Corrosion and odors in a wastewater collection system are mostly a function of the production and release of sulfide compounds, primarily hydrogen sulfide (H_2S). An evaluation of corrosion and odor requires an understanding of the types of compounds likely to cause problems and the mechanisms that form these compounds in a wastewater collection system.

9.3.1 Corrosion and Odor Compounds

Most odor-producing substances found in domestic wastewater are relatively volatile substances with molecular weights between 30 and 150. Most of these compounds form as a result of the anaerobic decomposition of organic material containing sulfur and nitrogen, which produces hydrogen sulfide, ammonia, carbon dioxide, and methane, among other compounds.

Hydrogen sulfide, with its characteristic rotten-egg smell, is the most prevalent and most notable odorous compound in the wastewater collection system. Hydrogen sulfide causes direct corrosion of metals such as iron, zinc, copper, lead, and cadmium. More importantly, hydrogen sulfide is oxidized to sulfuric acid, which can severely damage wastewater facilities.

Wastewater-pipeline corrosion is primarily microbially induced. This is a complex process involving many active organisms that, as is apparent from current research, is not fully understood or modeled. There are two distinct phases of microbially induced corrosion (MIC): 1) the conversion of ubiquitous sulfate in wastewater to sulfide, some of which is released as gaseous hydrogen sulfide, and 2) the conversion of hydrogen sulfide to sulfuric acid, which subsequently attacks susceptible pipeline materials.

This chapter briefly addresses the formation of sulfides in the collection system. For a more detailed exploration and for the formation of other odorous compounds, see Chapter 4.0.

Most sulfide is formed by bacteria living in a matrix of filamentous microbes and gelatinous material coating the submerged walls of wastewater pipes that is often referred to as the slime layer. The bacteria producing sulfide are strict anaerobes and, consequently, live beneath the water surface in gravity sewers and on the pipe walls in forcemains. The bacteria may also thrive in sludge and grit deposits found along the bottom of pipes. In order to produce sulfide compounds, the bacteria require a source of sulfur and a food supply. Sulfate, generally abundant in wastewater, is usually the common sulfur source, though other forms of sulfur, such as organic sulfur from animal wastes, can also be reduced to sulfide. The dissolved organic material prevalent in the wastewater provides an ample food supply for the bacteria to flourish. The sulfate reaction in a wastewater collection system can be described as follows:

$$SO_4^{2-} + \text{organic matter} + H_2O \xrightarrow{\text{bacteria}} 2HCO_3^- + H_2S \qquad (9.3.1\text{-}1)$$

The overall sulfide and sulfuric acid formation mechanism is illustrated in Figure 9-2.

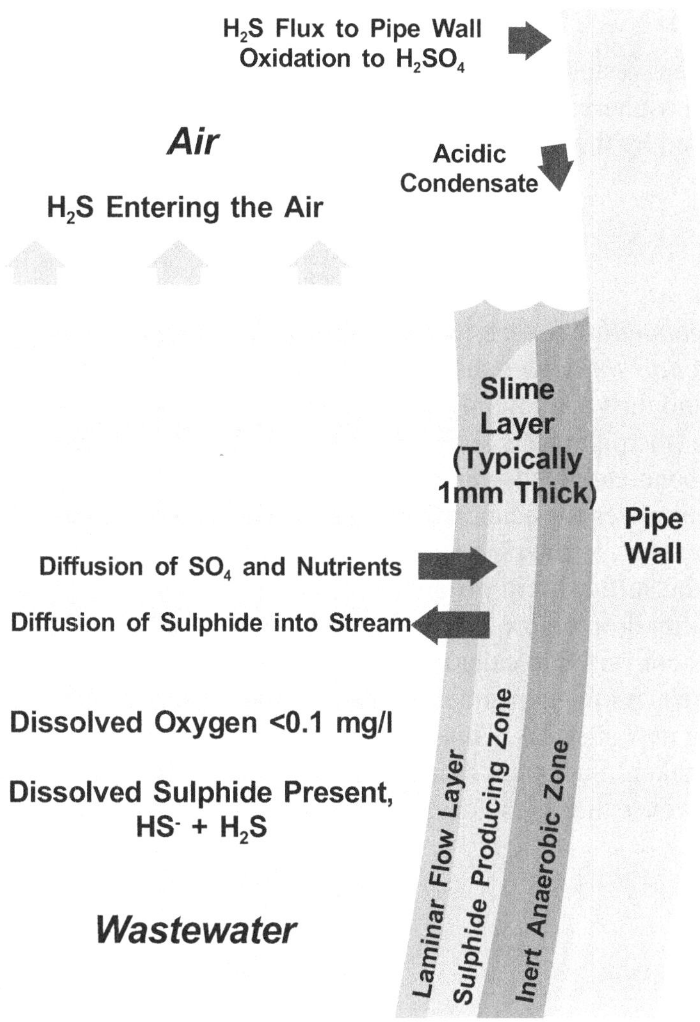

Figure 9-2. Overall Corrosion Mechanism in Wastewater Pipelines. (With permission from ASCE, 1989)

If concentrations of sulfate and dissolved organic material in the wastewater are high and if these materials are able to penetrate the solids deposits, then large amounts of sulfides can be produced. Many bacteria can reduce sulfate to sulfide, including the following:

- Assimilatory bacteria—those that assimilate inorganic sulfur and reduce it to sulfide within their protoplasm
- Proteolytic bacteria—those that can hydrolyze proteins and amino acids under anaerobic conditions

- Sulfate-reducing bacteria—specialized bacteria that use inorganic sulfate as the hydrogen acceptor in their energy cycle (*Desulfobacter* and *Desulfovibrio* are two such bacteria.)

The most troublesome problems occur when large amounts of sulfides are produced. Turbulence associated with normal flows in sewers encourages the release of hydrogen sulfide gas from the wastewater. Slugs of sulfides are often carried by wastewater down the interceptor, thereby providing a mechanism for significant hydrogen sulfide gas to be released. This slug effect complicates efforts to control hydrogen sulfide releases.

9.3.2 Hydrogen Sulfide Corrosion

Once sulfides are produced in the wastewater as the result of sulfate reduction, hydrogen sulfide gas will be released into the atmosphere. The hydrogen sulfide gas is oxidized on the pipe surface above the water line as described by the following equation:

$$H_2S + O_2 \xrightarrow{\text{bacteria}} H_2SO_4 \qquad (9.3.2\text{-}1)$$

Oxidizing bacteria, such as *Thiobacillus concretivorous, Thiobacillus neapolitanus, Thiobacillus ferro-oxidans,* or *Thiobacillus thioxidans,* as shown in Figure 9-3, create this second phase of corrosion. They are aerobic and thrive in low-pH environments. To grow, they require a source of sulfur (H_2S), a moist surface (the pipe walls above the water surface), and a carbon source (CO_2). The surface pH of new concrete pipe is generally between 11 and 13. Cement contains calcium hydroxide, which neutralizes the acids and inhibits formation of oxidizing bacteria when the concrete is new. However, as the pipe ages, the neutralizing capacity is consumed, the surface pH drops, and the sulfuric acid-producing bacteria become dominant. In active corrosion areas, the surface pH can drop below 2.0. As sulfides are formed and sulfuric acid is produced, calcium hydroxide is converted to calcium sulfate (gypsum). Calcium carbonate is easily eroded by wastewater. As the gypsum material is eroded, the concrete loses its binder and begins to spall, exposing new surfaces. This process will continue until the pipeline fails or corrective actions are taken. Sufficient moisture must be present for the sulfuric acid-producing bacteria to survive, however; if it is too dry, the bacteria will become desiccated, and corrosion will be less likely to occur.

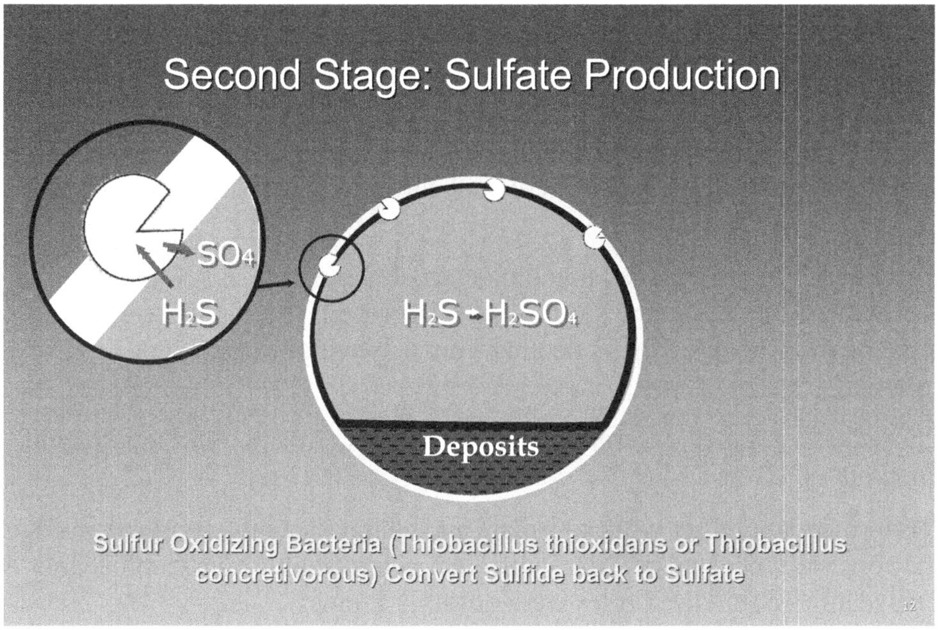

Figure 9-3. Second Phase of Corrosion: Sulfate Production.

Several factors are critical in determining how much hydrogen sulfide off-gassing will occur from the wastewater. Among the most important of these are the following:

- ***Dissolved sulfide concentration.*** Sum of liquid phase sulfide constituents in the wastewater $[H_2S + HS^-]$
- ***Wastewater pH.*** A lower pH results in a larger percentage of sulfide in the form of H_2S, which is then available for off-gassing. This is illustrated in Figure 9-4.

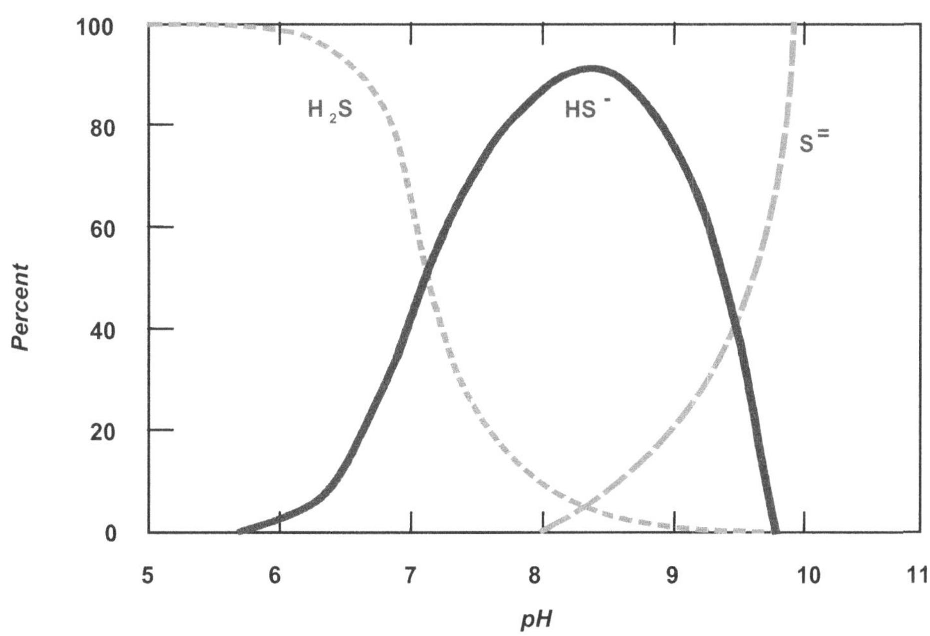

Figure 9-4. Sulfide Species as a Function of pH.

- ***Turbulence.*** Increased turbulence promotes off-gassing by providing a greater surface area for hydrogen sulfide to escape into the atmosphere and by bringing gases into more frequent contact with boundary layers.

Factors that play a less significant role include wastewater temperature, ambient hydrogen sulfide concentration in the sewer atmosphere, and barometric pressure of sewer atmosphere.

A relatively large number of articles was found on corrosion mechanisms. Most covered the two corrosion phases, sulfide formation and sulfuric acid formation. Among these are the following, grouped by their major themes. Papers dealing primarily with the corrosion mechanism were the following:

- Little et al. (2000) provided a detailed description of the various corrosion mechanisms and the biological sulfur cycle.
- Nica et al. (2000) performed laboratory work to isolate and characterize microorganisms that were involved in the corrosion mechanism. Many different heterotrophic, autotrophic, and fungi organisms were identified, but their roles were not fully known.
- Cho et al. (1994) isolated a new fungus, sporormia concretivora, that participates in corrosion. It was found to be acid resistant, thus accelerating the *Thiobacillus thiooxidans* corrosion mechanism.
- Hvitved-Jacobsen et al. (2000) developed a generation model that focused on sulfides. The model was used for design development of a gravity sewer in the River Emscher catchment area, in Germany.
- Vincke et al. (2001) analyzed the microbial communities on corroded sewer pipes using electrophoresis and found *Thiobacillus thiooxidans*, *Acidithiobacillus sp.*, and *Mycobacterium sp.* in the greatest numbers.
- Witzgall et al. (1990) performed a full-scale study of five collection systems and reported that whereas the sulfide formation correlated with the models, the corrosion did not.
- Weiss (1996) reported that sulfide generation rates in sewers in the County Sanitation District of Los Angeles ranged from 8 to 1,165 mg/m^2-hr in one system and from 137 to 257 mg/m^2-hr in another system. This resulted in the corrosion rates in concrete pipelines shown in Table 9-1.

Table 9-1. Concrete Pipeline Corrosion Rates.

	Corrosion Rate, in./yr	
H_2S Concentration, ppm	In Cast Pipe	In Spun Pipe
<1	<0.03	<0.02
1–3	0.03–0.05	0.02–0.03
3–8	0.05–0.08	0.03–0.05
>8	>0.08	>0.05

- Mori et al. reported corrosion rates in sewer pipelines in Japan of 5 mm per year (1991) and 4.3 to 4.7 mm per year (1992).
- In an unreferenced Belgian report, 2.3 to 3.1 cm of corrosion occurred after 6 years.

Eighteen other reviewed articles addressed corrosion mechanisms to a lesser extent.

Papers that discussed corrosion inhibitors included these:

- Bell et al. (2000) discussed the biochemistry and chemistry of the corrosion mechanism and found that ammonium salt in the concrete inhibited the bacterial growth of corrosion-causing microorganisms.
- Yamanaka (2002) speciated corrosive bacteria and also showed their neutralization by calcium formate; the laboratory tests showed that *Thioparus* is a precursor to *Thiobacillus thiooxidans*.
- Sima (1992) conducted inhibition experiments and showed that metabolic modifiers could eliminate acidophilic *thiobacilli*. Weak organic acids inhibited the *thiobacillus* mechanism. In particular, pyruvate and oxaloacetate acids were effective, but the reasons for this were not clear. This finding made unclear what the toxicity mechanism might be. It had been suggested that the modification of intracellular pH interfered with *thiobacillus* metabolism, and performing direct measurement of intracellular pH was suggested.

A general consensus of the reviewed literature was that measurable corrosion due to the hydrogen sulfide-sulfuric acid mechanism starts at hydrogen sulfide concentrations of 2-5 ppm.

9.4 Coatings

Coatings that are applied to the entire pipe circumference are distinguished from liners in that they are brushed or sprayed on rather than inserted as preformed material. Some coatings are applied during pipe manufacture, and others are applied in situ after installation, typically on pipelines that have started to exhibit corrosion. When perfectly applied, these coatings can be very effective, but each has potential shortcomings.

9.4.1 Factory-Applied

Factory-applied coatings include the following:

- Epoxy, typically applied to ductile iron, cast iron, and concrete pipes. Shortcomings include welding and cutting difficulties, potential for cracking during installation, and difficulty being repaired in the field.
- Concrete, typically applied to ductile or cast iron pipes. Shortcomings include the difficulty of welding or cutting, potential for cracking during installation, and the reduction in pipeline diameter.
- Galvanized coating, typically applied to steel pipes. Shortcomings include difficulty of welding and potential for accelerating corrosion.

Very few references to the above factory-applied coatings were found in the reviewed literature.

9.4.2 In Situ

Several papers reported on different in situ coatings, including coal tar epoxy, caustic soda, magnesium hydroxide, and organic acids. These coatings can be expected to need reapplication regularly, typically every few months to over a year. The issues with applying coatings include the logistics of accessing the sewers, traffic disruption, and the hazardous nature of chemical coatings, in particular, caustic soda, typically available at 25% and 50% solution.

In one instance, a magnesia slurry successfully reduced corrosion (Gebler et al., 2002) after a coal tar epoxy had failed. It was reported that the coal tar epoxy failed due both to hydrogen sulfide corrosion and to leaching by carbon disulfide.

Badia et al. (1992) reported that caustic crown spray, while raising pH from 2 to 11, provided 60 days of protection. Above pH 4, the corrosion rate was reduced from 0.3 inch per year to 0.03 inch per year. The life-cycle cost was substantially lower than that of PVC lining. In 1992 the costs were $11 per inch-diameter per linear foot for lining. The caustic crown spray costs in the same units were $0.03 per application. A typical application lasted 60 days, indicating an annual cost of $0.18 per inch-diameter per foot. For a 1-mile, 63-inch-diameter sewer, the cost comparison over 50 years was $3,700,000 in capital cost for sliplining versus $6,900,000 ($138,000 annually) for caustic crown spray.

Guan (2001) described many of the alternative coating materials for both the inside and the outside of pipes, including bituminous enamels, zinc/enamel, tape, liquid epoxy, cement mortar, heat-shrink sleeves, chlorinated rubber, and polyurethane.

Eight other reviewed articles addressed coatings.

9.5 Linings

Several types of plastic can be used as liners for sewer rehabilitation. Established limitations on the use of these materials are discussed below.

"Plastics" is a broad group of construction materials that can be divided into two major categories: thermoplastics and thermosetting plastics.

Thermoplastics soften and flow when heated. They are formed by joining individual carbon atoms into linear chains that create very large molecules. The most common thermoplastic sewer pipe materials are PVC and polyethylene. Others are polystyrene, polypropylene, acrylonitrile butadiene styrene, acrylics, nylons, polycarbonates, and thermoplastic polyesters.

Thermosetting plastics are formed as networked, rather than linear, structures. The network is formed by heating the base materials together. Unlike thermoplastics, once formed, thermosetting plastics do not melt. Thermosetting plastics are used for rehabilitating cured-in-place pipe (CIPP) and are typically composed of thermosetting polyesters and epoxies. Other thermosetting materials are phenolics, amino resins, and silicones.

Thermoplastics are more easily formed (and deformed) than thermosetting plastics, which are more rigid and have a much higher modulus of elasticity.

Table 9-2 summarizes the alternative liner types.

Table 9-2. Pipeline Lining Rehabilitation Alternatives.

Rehabilitation Option	Products	Pipe and Liner Materials	Available Diameters (in.)	Structural Issues	Approximate Hydraulic Capacity Reduction
Sliplining with continuous pipe	Chevron/Plexco, Phillips/ Driscopipe	HDPE fusion-welded	4–54	Structural repair, wall thickness based on condition of host pipe. Annular space grouting results in a composite pipe	40%t reduction for 32-in. outside diameter, 28.2-in. inside diameter, SDR 17 inside a 36-in. pipe

Rehabilitation Option	Products	Pipe and Liner Materials	Available Diameters (in.)	Structural Issues	Approximate Hydraulic Capacity Reduction
Sliplining with segmented pipe	Chevron/ Spirolite	HDPE	18–120	Structural repair, wall thickness based on condition of host pipe. Annular space grouting results in a composite pipe	46% reduction for 27-in. inside diameter, 31-in. outside diameter, Class 160 inside a 36-in. pipe
	Lamson Vylon	PVC	21–54	Structural repair. Annular space grouting results in a composite pipe.	29% reduction for 30-in. inside diameter, 32.2-in. outside diameter inside a 36-in. pipe.
	Hobas	Reinforced thermosetting resin pipe	4–144	Structural repair, wall thickness based on condition of host pipe. Annular space grouting results in a composite pipe	24% reduction for 31-in. inside diameter, 33-in. outside diameter (bell) inside a 36-in. pipe, (smaller annulus okay due to high allowable jacking force). Prone to failures if stressed by bending
CIPP	Insituform, FirstLinerUSA, CIPP Corp., National Liner, InLiner, Impreline Tech, Permaliner, Cure-line	Nonwoven, multi-layered felt liner with thermosetting resin (isophthalic polyester, vinyl ester, or epoxy)	4–108; noncircular sections okay	Structural repair, wall thickness, and resin selection based on condition of host pipe	4% increase for 0.67-in.-thick liner inside 36-in. pipe. Based on 10 foot depth, H-20 live load, 5% ovality, and computed with formulas from ASTM F-1216
Fold and form, deformed and reformed, die-draw liners	AM-Liner, NuPipe, EX Method, Ultraliner, U-Liner, Sure-Line, Rolldown, Swagelining	PVC, or PVC alloy, PE	4–24 maximum for selected products	Structural repair, wall thickness based on condition of host pipe	Reduction not applicable for greater than 24-in. pipes since product is not made in those sizes
Site-folding PE pipe	PIM Corp Subline	PE	3–60	Nonstructural repair	1% reduction for 36-in. outside diameter, 34-in. inside diameter
HDPE liner with grout	Trolining	Studded HDPE sheeting with grouted annulus	8–95	Structural repair, wall thickness and optional use of reinforcing based on condition of host pipe. Annular space grouting results in a composite pipe	5% increase for maximum 0.63-in. wall thickness inside 36-in. pipe
Spiral-wound PVC liner with grout	Danby, Ameron	PVC strip with grouted annulus	24–120	Structural repair, wall thickness and grout characteristics based on condition of host pipe. Annular space grouting results in a composite pipe	8% reduction for 33-in. inside diameter inside 36-in

Rehabilitation Option	Products	Pipe and Liner Materials	Available Diameters (in.)	Structural Issues	Approximate Hydraulic Capacity Reduction
PVC sheet liner with grout	Linabond, Riblock	PVC sheeting with thermosetting polymer in annulus	48 and larger	Structural repair, thickness of polymer based on condition of host pipe	1% reduction for 1-in. thickness applied to 36-in. pipe (thickness of 1/4 in. minimum to 6 in. or more available)

Table 9-3 matches criteria with recommended lining rehabilitation suggestions.

Table 9-3. Pipeline Lining Criteria.

Criterion	Suggestion
Corrosion and abrasion resistance	Advanced hydrogen sulfide corrosion is the principal reason that rehabilitation of these pipelines is necessary. Materials used for rehabilitation must have superior resistance to sulfuric acid. Resistance to other chemicals must be considered for pipelines near significant industrial discharges. Abrasion or erosion of the pipeline invert by sand, grit, and other solids in sewage is also a concern; thus the selected materials must have high resistance to abrasion.
Temperature sensitivity	Due to extreme daytime temperatures in certain hot climates, some sewer and street construction projects are scheduled for nighttime construction. Depending on the time of year these projects are bid and constructed, construction methods and procedures that are able to withstand high temperatures throughout installation, as well as warm ground conditions, may become an issue. In addition, materials with high coefficients of thermal expansion can create problems, such as blocking or shearing off of lateral connections, if exposed to a wide temperature range. High density polyethylene pipe is a rehabilitation material that is susceptible to this problem.
Lateral reinstatement	The reinstatement of sewer laterals is normally completed either internally, using a robotic cutter guided by a television camera, or externally, by excavation and replacement. External reinstatement is preferable in those instances where existing connections are in poor condition. However, internal connection, if appropriate, results in less disruption to the project area.
Bypass pumping requirements	Many rehabilitation methods require full or partial bypass pumping of sewage flows throughout the installation process. Bypass pumping can be a significant portion of the total rehabilitation project cost (upward of 35%; City of Phoenix, 2005). Bypass pumping requires round-the-clock pumping equipment, maintenance personnel, standby pumps and generators, and long runs of surface piping that allows traffic to cross at driveways and intersections.
Disruption	The more excavation required for rehabilitation, the greater the disruption to traffic, businesses, and neighborhoods. For projects located under major streets or other critical surface improvements, minimizing excavation can be an important factor in selection of methods. Noise and odor can create additional impacts from the project.
Quality control	For a rehabilitation method to be successful, it must be properly installed and inspected. Some methods are more installation sensitive than others, and have a higher risk of potential problems. These problems may be apparent in the short term or may not show up until later. Inspection by a knowledgeable Resident Engineer, including sample testing of the installed product, is critical to project success.

Criterion	Suggestion
Track record/estimated service life	The sewer rehabilitation industry has evolved rapidly in recent years, with many new entrants into the market. Some of these ventures have been short lived for a variety of reasons. A proven history of successful installation and performance is an important predictor of a product's likelihood for success on future projects. The estimated service life for each rehabilitation method is a key factor in assessing the value of the various alternatives.
Relative cost	Cost is always a primary consideration in selecting pipeline rehabilitation methods. The most expensive technologies often serve specific market niches that make them the only appropriate method for certain circumstances. However, a simple comparison of relative costs is usually sufficient for screening purposes, without the need for detailed cost estimates, in those situations where a less expensive method will result in an equivalent final product.
Other construction issues	This criterion is included to identify any other important factors not previously covered and typically unique to a given rehabilitation technique. Each of the pipelines to be rehabilitated has a sanitary flow that must be considered during the design and construction phases of the project. Traffic control plans will be required from the Contractor. Insertion pits or trenches should be within rights-of-way and located to avoid street intersections and access driveways. Sites are frequently located in residential and commercial/light industrial areas; access to some residences and/or businesses may be blocked for a brief period of time. During sewer rehabilitation there will be odor generation and the potential for release of these odors to the atmosphere.

The reviewed literature described several full-scale lining installations, and several provided cost information. Uhren and Gilbert (2001) evaluated various pipe materials, concrete additives, and liners for a new sewer tunnel in Columbus, Ohio, and found that lined concrete pipe was the best option for that specific case. The exact lining method was to be determined during detailed design.

Saricimen et al. (2003) reported on 24-month laboratory tests of two proprietary, high-alumina, cementitious linings and found that one, "SC," performed significantly better. Both linings performed better than Type 1 cement with 20% fly ash cement and better than Type 1 cement with 10% silica fume. Accelerated laboratory tests and a two-year field study were performed for a collection system in Jubail, Saudi Arabia. It was recommended that the field study be continued.

Los Angeles County in 1995 conducted pilot tests of polyester mortar coatings and PVC liners and determined that 19 of 71 methods analyzed were successful. They have tested seven coatings at full scale since 1983. Comparative cost information is provided, showing generally that coatings have been significantly less expensive than PVC liners.

Eight other reviewed articles addressed linings.

9.6 Corrosion-Resistant Materials

Among many possible wastewater pipeline materials, some of the most common are the following:

- ***Vitrified clay.*** Completely resistant to H_2SO_4 corrosion, but joining materials such as mortars can corrode it

- ***Concrete pipe.*** High corrosion potential. Can be constructed of mixtures that improve resistance
- ***Steel pipe.*** High corrosion potential
- ***Cast iron pipe.*** High corrosion potential but lower than that of steel
- ***Ductile iron pipe.*** Better resistance than steel or cast iron. Carbon and silicon content increases resistance.
- ***Plastic pipe.*** Many plastic pipe materials available, including PVC, polyethylene, high-density polyethylene, and fiber-reinforced plastic. Highly resistant to corrosion, but varying structural properties.

Of the 16 reviewed articles on corrosion-resistant materials, 10 addressed additives to concrete.

Bell et al. (2000) reported on the use of an ammonium salt as a concrete additive and its subsequent corrosion reduction.

Vroom et al. (2000) compared sulfur concrete and hydraulic cement concrete, concluding that sulfur concrete was better structurally and in terms of corrosion resistance. Shortly thereafter, Wahshat (2001) performed pilot tests of sulfur mortar and polymer-modified sulfur mortar and found that the latter provided good corrosion resistance.

A laboratory test reported by Yamanaka et al. (2002) showed that the addition of calcium formate to the concrete mixture completely inhibited the growth of sulfur-oxidizing and iron-oxidizing bacteria.

Research into polymer concretes performed at the Catholic University of Leuven, in Belgium (Beeldens et al., 2001), showed that most polymer additives improved corrosion resistance and increased flexural strength but decreased compressive strength. However, two polymers, polyacrylic and styrene-acrylic, exhibited weaker structural properties.

As part of research to determine best test methods for corrosion measurement, De Belie et al. (2004) performed an immersion-drying-brushing test cycle and found that limestone aggregate provided good corrosion resistance and that lower water-to-concrete-ratio cements had better resistance; they suggested that water absorption properties be included in the Pomeroy corrosion model.

Hewayde et al. (2003) performed experimental work on selected admixtures, concluding that Metakaolin improved resistance and strength and that corrosion inhibitors in general improved resistance but decreased strength. Contrary to De Belie et al.'s (2004) results, reported above, Hewayde et al. reported that *higher* water-to-concrete ratios provided better resistance but decreased strength.

Monteny et al. (2001) reported that styrene-acrylic ester polymer had the best resistance of five tested additives, whereas acrylic and styrene-butadiene polymers decreased corrosion resistance.

Similar polymer concrete research by Vincke et al. (2002) at the University of Ghent, in Belgium, showed that styrene acrylic ester polymer exhibited increased resistance but that acrylic polymer or silica fume provided less resistance. Their work showed that neither vinyl copolymer nor styrene-butadiene polymer affected corrosion resistance.

Seven other reviewed articles addressed corrosion resistant materials.

9.7 Hydraulic Design

Most rehabilitation methods reduce the interior cross-sectional area of the pipeline and typically change the pipeline wall surface roughness. These factors impact the hydraulic capacity of the sewer. Where capacity is limited, a loss of cross-sectional area may render some rehabilitation methods unacceptable. Alternatives for pipeline rehabilitation must maintain sufficient hydraulic capacity for the existing and projected future flow conditions, including stormwater inflows. In evaluating rehabilitation techniques, the associated reduction in hydraulic capacity of a particular method must be compared to the estimated future flow rates for each specific project.

A major factor in selecting a rehabilitation technique is the available hydraulic capacity of the sewer following rehabilitation. The rehabilitated sewer must be capable of conveying the maximum daily flows with a reasonable factor of safety to allow for uncertainties concerning actual future population growth, winter flow increases, and stormwater inflows. After rehabilitation, a reserve capacity is recommended to account for these uncertainties. Minimum velocities at full flow (d/D = 1.0) will not control the design because rehabilitation of the sewers will not decrease the velocities from their current values.

9.8 Cost Comparison of Alternative Measures

The review of documents describing pipe corrosion protection methods produced over 100 coating and lining techniques. Conspicuous by its absence, however, was cost data on these techniques. Only a handful of papers included cost data.

While it is recognized that costs are site specific and that labor can be a high proportion of the cost component, comparative cost data should be obtainable for typical gravity pipe systems. Needed too is a comparison of the costs of coating and lining techniques with costs of the other common protection technique, chemical addition.

9.9 Rehabilitation Scheduling and Funding

It is common practice to plan for repairing and replacing corroded sewer pipelines and to do so, ideally, on the basis of condition assessments. A sewer replacement project investment period can be from 50 to 100 years. The ranking of replacement projects is increasingly performed using a risk-based approach in which the consequences of failure are weighed against the cost of replacement. Those consequences might include property damage, neighborhood and traffic disruption, and odor and health impacts.

The funding source for replacement projects is inevitably the agencies' customers, who must be convinced of the need to replace sewers by having the consequences of failure explained to them. Under normal circumstances, grants are rarely available for sewer replacement.

9.10 Research Gaps

Guidance is needed for agencies having to provide pipeline corrosion protection for both new and existing installations. Though many vendors may claim that their corrosion protection techniques are the most effective, objective data are needed.

For new pipes, the alternatives available include corrosion resistant pipe, linings, and coatings. For existing pipes, the alternatives include linings or coatings.

In addition, the above pipe material corrosion resistance techniques should be compared to other techniques, such as chemical addition.

9.10.1 Objective

To provide agencies with guidelines and a decision path to enable them to determine most effective corrosion protection for new and existing gravity sewer pipelines. Produce a guidance document for cost-effective corrosion protection of gravity pipelines whose diameters range from 24 to 84 inches.

9.10.2 Program

This study would consist of several phases:

9.10.2.1 Phase 1—Conduct Expanded Literature Search

In addition to using the papers discovered in the current project, search for other papers online and by contacting public agencies, consultants, vendors, and contractors.

9.10.2.2 Phase 2—Determine Costs of Pipe Corrosion Protection Techniques

Using information from Phase 1 and by contacting vendors, develop cost ranges for alternative pipe-coating techniques.

9.10.2.3 Phase 3—Compare Pipe Lining to Chemical Addition

Compare the costs developed in Phase 2 with the costs developed in another research gap topic, "Life Cycle Analysis of Chemical Addition on Downstream Processes."

9.10.2.4 Phase 4—Report Findings

Produce a report summarizing findings and including guidelines as to which corrosion protection techniques are cost effective under what conditions.

CHAPTER 10.0

RESEARCH GAPS EVALUATION AND SUMMARY

This chapter summarizes the initial 13 research gaps identified during the Phase 1 literature search and provides a ranking for those research gap areas determined to be highest priority based upon input from WERF and WERF subscribers. The WERF research team developed gap summaries providing "problem statements" to define each gap along with potential "research approaches" geared to fill each data gap. The research gaps are listed below and discussed in the following pages:

1. Better definition of sewer ventilation so that natural air movement and displacement in collection systems may be better understood.
2. Better definition of the odorous and corrosive compounds other than H_2S found in collection systems.
3. Better definition of the treatment effectiveness for non-hydrogen sulfide odorous compounds by means of available liquid and gas phase treatment systems.
4. Improved predictive relationship between hydrogen sulfide (H_2S) gas concentration and corrosion rates in sewer systems.
5. Improved understanding of the impacts of collection systems chemical dosing on the downstream treatment plants.
6. Characterization of the importance of substrates on sulfide generation rates in collection systems.
7. Definition of the feasibility, practicality, and related costs of using oxygen treatment in gravity sewers.
8. Improved understanding of available biological treatment using enzymes and bacterial cultures.
9. Documentation of the mechanisms for odor and corrosion control using magnesium hydroxide pH adjustment.
10. Development of design guidance for sizing air jumpers and related ventilation systems. (This research gap may become part of the effort associated with Gap 1.)
11. Improved understanding of the relative effectiveness of biotechnology-based odor control systems in cold climates.
12. Better understanding of the unique odor and corrosion impacts from fats, oils, and grease.
13. Clear definition of the odor and corrosion effect of using sewers as reactor vessels to achieve partial wastewater treatment. (This practice is gaining acceptance in Europe and may need to become better understood worldwide.)

Gaps 1 and 2 were initially ranked very closely by the project research team. Based upon initial review with the WERF PSC, the following six research gap areas were identified as high-priority during a meeting with the PSC at the Detroit Collections Systems Conference in August 2006. A WERF subscriber survey requesting that WERF members provide feedback and ranking for the following six gap areas was then completed:

- Improved definition of sewer ventilation to better understand natural air movement and displacement in collection systems.
- Better define the odorous and corrosive compounds found in collection systems. (What are the non-hydrogen sulfide odor- and corrosion-causing compounds? This gap will include evaluation of a representative variety of collection systems including various climates, CSO and non-CSO systems, and those influenced by fat oil and grease impacts.)
- Better definition of the treatment effectiveness for non-hydrogen sulfide odorous compounds.
- Better understanding of the relationship between hydrogen sulfide gas concentrations and corrosion rates.
- Better understanding of the practicality and related costs of using oxygen treatment in gravity sewers.
- Clear definition of the odor and corrosion effect of using sewers as reactor vessels to achieve partial wastewater treatment.

The subscriber ranking survey was issued to WERF subscribers in October 2006 in order to assist in the final ranking of the highest-priority research gaps. On the basis of this survey, the following were identified as the top-ranking research gaps:

1. Improved definition and understanding of natural sewer ventilation is needed to better understand air movement and displacement in collection systems and the related impacts on odor and corrosion.
2. Improved understanding of the effects of hydrogen sulfide gas concentrations and the relationship to concrete corrosion rates in collection systems is needed.
3. Better definition of the non-hydrogen sulfide odor- and corrosion-causing compounds found in collection systems.
4. Better definition of the treatment effectiveness for non-hydrogen sulfide odorous compounds.

These four research gaps are considered to be of the highest priority. It is anticipated that Phase 2 of this research program will be geared towards addressing these research gap areas in as much as possible. Where additional effort and budget beyond the scope of this project is needed, alternative collaborative research opportunities will be pursued.

The remainder of this chapter outlines each of the initially identified 13 research gap areas. The top four include initial budgetary estimates.

10.1 Natural Ventilation of Sewers

The nature and extent of air exchange (ventilation) between the ambient atmosphere and headspace of sewers is perhaps the most important variable affecting the generation and release of odorous and corrosive gases in gravity sewers. Ventilation replenishes the oxygen in the headspace of sewers, affects the degree of condensation on sewer crowns (and hence sewer corrosion), and releases and conveys odorous gases from the sewer system.

10.1.1 Problem Statement

Despite the importance of sewer ventilation, little progress has been made in fundamentally understanding the factors that affect natural ventilation, how those factors can be controlled to optimize ventilation in corrosive and odorous sewer environments, and how to estimate, even by rough approximation, how sewer ventilation rates and patterns will change with modifications to sewer systems. The following is an overview of some research conducted on this topic.

Pescod and Price (1981) performed lab-scale experiments involving a 0.3-m sewer operating at various depths and rates of flow and measured average headspace air velocities. Their work focused primarily on liquid-drag induced natural ventilation. Parker and Ryan (2001) showed how inert tracers such as carbon monoxide can provide information on air flows in specific sewer reaches, but the technique has not been applied to a wide range of reaches thus allowing for more generalized knowledge. Edwina-Bonsu and Steffler (2004) developed a state-of-the-art model of fluid mechanics in gravity sewers, but focused entirely on the effects of wastewater drag and employed a fairly sparse data base in evaluation of the model. Olson et al. (1997a, 1997b) made novel use of the first law of thermodynamics to predict air flow in sewers given competing ventilation mechanisms, and performed small-scale sewer experiments that correlated very well with model predictions. However, their model was never tested with actual field data. Improved knowledge in this area is critical to advancing integrated solutions to sewer odor and corrosion problems.

10.1.2 Proposed Research Gap Evaluation Method

10.1.2.1 Objective

The objective of the proposed research is to 1) develop a database of field tracer data that can be used to evaluate existing and future models related to sewer ventilation; 2) evaluate existing models for the prediction of sewer ventilation; and 3) modify, as appropriate, the existing models or develop new models based on the results of step 2.

10.1.2.2 Proposed Program

This study would consist of the phases described below.

Phase 1—Identification of Sewer Reaches During this phase, sewer components and locations for testing will be identified. In addition, sampling and analytical procedures will be developed. For the purpose of this proposal, it is assumed that up to eight sampling events will be completed. As part of the sampling plan, a list of parameters necessary for applying computational fluid dynamic models (such as Edwina-Bonsu and Steffler [2004]) and thermodynamic models (such as Olson et al. [1997a, b]) for predicting ventilations (rates and directions) and relationships developed by Pescod and Price (1981) will be identified.

The following activities will be completed during Phase 1:

- Identification of sewer components/operating conditions to be investigated. Potential components might include: 1) inlets to siphons or wet wells; 2) drop structures, both conventional and vortex design; 3) gravity sewers, with and without sidestream connections; and 4) warm and cold climate conditions (where liquid-air thermal gradients may affect out-gassing)

- Identification of potential testing locations. Utility partners to be solicited for support may include Orange County (Calif.) Sanitation District, Los Angeles County Sanitation District, and King County (Wash.)
- Review of candidate location system physical information
- Site visits to finalize ventilation testing locations
- Sampling and analytical plan development

Phase 2—Field Studies Field studies will be completed in the reaches identified in Phase 1 using inert tracer methods similar to those of Parker and Ryan (2001). Data will be collected as additional input to models, such as wastewater and headspace temperatures, and wastewater flow rates during gas tracer experiments. The intent of this testing is to obtain sufficient data to characterize and define various operating conditions or parameters and the resultant natural ventilation rates and to allow assessment of the accuracy of existing ventilation models. Sampling will consist of the following tasks:

- Use inert tracers to quantify ventilation rates
- Monitor wastewater flow and temperature
- Monitor headspace temperature and pressure
- Measure liquid phase sulfide concentration (grab sampled using adsorbent tubes)
- Measure vapor phase hydrogen sulfide concentration (continuous sampling using hydrogen sulfide OdaLogs)

It is expected that each sewer component can be tested in a day. At least two utility partners will be used, one to test during warm conditions, the other during winter conditions:

- Utility Partner A—warm climate testing × 4 component tests = 4 tests
- Utility Partner B—cold climate testing × 4 component tests = 4 tests

Phase 3—Existing Models Evaluation The ability of existing sewer ventilation models to accurately predict ventilation rates will be evaluated using the data collected in Phases 1 and 2. For the purpose of this proposal, it is assumed that up to three models will be compared: for example, the Pescod and Price (1981) air velocity-wastewater velocity relationships, the Edwina-Bonsu and Steffler (2004) computational fluid dynamics-based model, and the Olson et al. (1997a, b) thermodynamic-based model. Where possible, recommendations on improved methods for estimated natural ventilation rates will be suggested.

Phase 4—Report Preparation A draft report will be prepared documenting activities completed during Phases 1 through 3. The report will include:

- Documentation of WCS components and locations selected for testing
- Sampling and analysis plan
- Database of testing results
- Comparison of field-measured versus model-predicted results
- Recommendations on potential improvements to natural ventilation rate algorithms

10.1.3 Schedule

For the purpose of scheduling, it is assumed that notice to proceed for this research activity will be issued by May 1, 2007.

Table 10-1. Natural Ventilation of Sewers: Schedule.

	Start Date	End Date
Phase 1	05/01/2007	05/01/2007
Phase 2	05/01/2007	09/01/2007
Phase 3	09/01/2007	12/01/2007
Phase 4	12/01/2007	02/28/2008

10.1.4 Budget

Table 10-2. Natural Ventilation of Sewers: Budget.

	Total ($)
Phase 1	67,595
Phase 2	101,989
Phase 3	21,902
Phase 4	21,245
Total	212,731

10.2 Relationship between Sewer Corrosion and Headspace Hydrogen Sulfide Concentration

Sewer system owners often use modeling to predict sulfide transport within sewers and emissions from sewer headspaces. Some models also predict corrosion and concrete component life based on liquid phase sulfide. Such predictions rely on indirect parametric relationships and provide only order-of-magnitude accuracy.

10.2.1 Problem Statement

Sewer component corrosion is known to be strongly correlated to headspace hydrogen sulfide concentrations, a parameter that is easily measured in the field. Improved knowledge of this relationship would enable more accurate modeling that could be used to better predict pipe life and identify potential corrosion spots. Yet, the relationship between *headspace* hydrogen sulfide and corrosion has not so far been well characterized.

Several studies that characterize the bacterial conversion of hydrogen sulfide to sulfuric acid and subsequent corrosion have been completed. Findings from these studies could be incorporated with an investigation of mass transfer of hydrogen sulfide from the bulk gas phase to the damp film coating unsubmerged pipe walls. The study should track parameters that influence mass transfer, such as headspace ventilation, air temperature, and water level. By combining existing knowledge of bacterial uptake of hydrogen sulfide and subsequent corrosion with new knowledge of mass transfer, a model algorithm could be developed or old algorithms improved.

10.2.2 Proposed Research Gap Evaluation Method

10.2.2.1 Objectives

The objectives of the proposed study are the following:

- Identify the rate of bacterial uptake of hydrogen sulfide on unsubmerged pipe walls and subsequent corrosion in the literature.

- Develop a mathematical model to characterized mass transfer between the bulk headspace air and the un-submerged pipe wall film.
- Determine experimentally the hydrogen sulfide deposition under varying conditions.
- Combine the results into an algorithm that relates corrosion with hydrogen sulfide gas concentration.

10.2.2.2 Program

The program of research would consist of three phases.

Phase 1—Mathematical Model Development Phase 1 will be implemented to describe mathematically the variables, reactions, mass transfer processes present between the bulk hydrogen sulfide gas phase and concrete corrosion of un-submerged pipe walls. Reaction rate constants and mass transfer coefficients will be located in the technical literature, where available. Three months will be budgeted to complete Phase 1.

Phase 2—Experimental Determination of Hydrogen Sulfide Deposition Phase 2 will be completed in a laboratory where important variables can be controlled. It is proposed that experiments be conducted in a horizontal cylindrical chamber partially filled with water. The water need not be flowing. However, the water level should be variable. Also, headspace ventilation and hydrogen sulfide gas concentration should be variable. Five months will be budgeted for the completion of Phase 2.

Phase 3—Algorithm Development During Phase 3 the results of Phases 1 and 2 will be combined into an algorithm that can be used to predict pipe corrosion based on headspace hydrogen sulfide gas concentration. Four months will be budgeted for Phase 3, including documenting Phases 1, 2, and 3 in a final report.

10.2.3 Budget

Table 10-3 summarizes the schedule and budget.

Table 10-3. Relationship between Sewer Corrosion and Headspace Hydrogen Sulfide Concentration: Schedule and Budget

	Schedule	Budget ($)
Phase 1	05/2007–08/2007	25,748
Phase 2	08/07–01/08	128,466
Phase 3	01/08–05/08	42,771
Total	12 months	196,985

The approach outlined above does not include field verification. If an extensive field verification step is added then it is estimated that additional time and cost would be incurred. Total time might increase to as much as two years and estimated total costs could approach $379,000.

10.3 Compounds in Sewer Gas That Lead to Odor Complaints and Corrosion: Non-H_2S Odor- and Corrosion-Causing Compounds

10.3.1 Problem Statement

Odors associated with wastewater collection systems typically are attributed to the presence of hydrogen sulfide. Thus many technologies used to control sewer odors are directed toward suppression of hydrogen sulfide generation (for example nitrate addition to wastewater), or to its destruction/removal following generation (ferrous chloride addition to wastewater or treatment of headspace gases in biofilters). Conditions in sewer systems can produce other compounds of low odor threshold, such as those listed in Table 10-4. The contribution of such compounds to odor nuisance complaints is not well understood, and little has been done to study this odor phenomena. Collection systems may well have locations where effective control and removal of hydrogen sulfide will not lessen an odor problem. A control strategy may need to be developed for other odorous compounds besides hydrogen sulfide to alleviate a collection system odor problem.

Table 10-4. Odorous Compounds Associated with Untreated Wastewater.

Class of Compound	Relevant Examples
Inorganic gases	Ammonia, hydrogen sulfide
Mercaptans	(Allyl; amyl; benzyl; crotyl; ethyl; methyl; propyl) mercaptans
Other organic sulfur	Dimethyl sulfide, diphenyl sulfide, thiocresol, thiophenol
Amines	(Dibutyl; diisopropyl; dimethyl; ethyl; methyl; triethyl) amines
Diamines	Cadaverine (1,5-pentanediamine); putrescine (butanediamine)
Other organic nitrogen	Indole, pyridine, skatole
Volatile fatty acids	(Acetic; propionic; butyric; isovaleric) acids

10.3.2 Study Goals and Objectives

The following study goals and objectives for the identification of compounds that may contribute to collection system odors:

- Survey several collection systems throughout the U.S. and in Australia to evaluate locations that may have odors associated with compounds that are not hydrogen sulfide.
- Identify the most significant compounds, in addition to hydrogen sulfide, that contribute to perceived odors in sewer headspace gases. Due to limitations in existing analytical methods the contribution to odor by amines, organic nitrogen compounds and volatile fatty acids will have limited consideration during this study.
- Evaluate the contribution of odorous compounds to perceived odors relative to the contribution from hydrogen sulfide.
- Develop guidelines or procedures for distinguishing between locations where control of hydrogen sulfide alone is sufficient to minimize odors and locations where odor control must target control of other compounds.

10.3.3 Experimental Approach Proposed Research Gap Evaluation Method

A gas chromatograph mass spectrometer olfactometer (GC/MS/O) will be used. The WERF team has access to two of these systems, one in Australia at the University of South Wales and another in Los Angeles through the partnered labs of Los Angeles County Sanitation District.

In the GC/MS/O, individual odorous compounds in the sample gas are resolved using gas chromatography. A gas sample is introduced into a gas chromatograph and cryo-trapped onto a column. As different compounds are eluted from the gas chromatograph column they pass through a flow splitter. A portion goes to the mass spectrometer, where a spectral signature is generated that enables identification and quantitation of each analyte. Another portion is directed to a sniffing port that presents the gas to a trained chemist. For each compound that registers an olfactory response, the odor is characterized and ranked on an odor intensity scale containing up to seven steps. The GC/MS/O set-up enables the spectral signature produced by each analyte in the mass spectrometer to be correlated to the odor perceived by the human subject. The correlation is dependent on the gas chromatograph methodology, which will impart a slight difference between the mass spectrometer spectra and the odorgram.

The approach to characterizing odors in sewer gas will include the following tasks:

- Establish protocols to ensure consistency in sample collection and proper handling of equipment.
- Screen sampling locations.
- Select sample locations for detailed study.
- Conduct sampling and analysis.
- Conduct data analysis and prepare a report.

10.3.3 Sampling Protocols

Sampling collection Grab samples for GC/MS/O analysis will be collected in either fused-silica-lined stainless steel canisters 6 L in volume or Tedlar bags (whichever is most appropriate). Samples analyzed for odor strength will be collected in commercially produced Tedlar bags that have been condition with high purity air to remove impurities. Sampling depth—All samples to be collected from inside the collection systems below the manhole covers.

Holding times Odor panel samples that are collected in Tedlar bags will be analyzed within 24 hours of collection. Samples that are collect stainless steel canisters can be held for longer periods of time because of the nonreactive nature of the canisters but analyses will typically be made within 24 to 48 hours after collection of sample.

10.3.4 Selection of Sampling Locations

To determine the influences that different wastewater collection systems have on generation of odorous compounds in sewer headspaces, samples will be collected at several different collection systems. The sample locations will be situated in different regions of the U.S. so that regional climatic differences can be compared. In addition, the sample locations will include large and medium collection systems, systems with high and low percentages of industrial waste, systems with and without force mains and will include older combined sewer overflow systems and newer non-combined sewer overflow systems.

The proposed U.S. locations for sampling include the following:

Table 10-5. Collection Sampling Locations and Their Properties.

Collection System	Collection System Properties
Los Angeles County Sanitation Districts	Large collection system in a warm climate with significant industrial waste
Orange County (Calif.) Sanitation Districts	Large collection system in a warm climate with little industrial waste
King County (Wash.)	Large system in a cool wet climate, with little industrial waste and a mixture of force mains and gravity sewers
Hampton Roads (Va.) Sanitary District	Medium-sized east coast system with a significant number of force mains
Chicago Metropolitan Water Reclamation District	Large system with a major industrial waste component
Pittsburgh Water Authority	A medium-sized system with combined sewer overflows

Locations are also being identified in Australia. Other collection systems could also be included to try and compare additional field conditions that could influence the generation of odorous compounds. Final selection of collection sites will attempt to balance locations where hydrogen sulfide is the predominant contributor to odor character against locations where other compounds may exist at significant levels. As an initial step, utility partners will be asked to submit a list of candidate locations from their collection systems where odor control has been difficult. As part of the submittal, the utility partner will include basic information for each candidate location such as range of flows, sewer diameter, depth to sewer invert, range of flow depth, type of manhole structure, sewer slope, distance to treatment plant, type of wastewater (for example, mostly residential, mixture of industrial and residential), identification of significant industrial discharger and type of industry, data, odor characteristics, dissolved sulfide, total sulfide, hydrogen sulfide). The utility will also briefly explain why each location is proposed for further evaluation.

Candidate locations will be selected for preliminary evaluation. Gas samples from those locations will be presented to an odor panel for initial odor characterization. Each location could be sampled two times. During sample workup, the gas will be split into two duplicate samples.

10.3.5 Laboratory Methods

10.3.5.1 Odor Measurements

Odor level in sewer headspace gas samples will determined by a certified odor panel at the Los Angeles County Sanitation District Joint Water Pollution Control Plant laboratory. The panel can measure up to about six samples per day. (Above six, fatigue of the senses tends to occur.)

Panel members will rate the odor intensity of the undiluted sample on a seven-point scale and assign descriptors for the odor chosen from a defined list (for example, rotten egg, rancid, fermented cabbage, earthy).

10.3.5.2 GC-MS–O

Samples will be analyzed at both the Los Angeles County Sanitation District and the University of South Wales labs. Each lab will spend roughly 12 weeks running two tests per week. As such, no more than 48 samples are expected.

- Tasks 1: Initial survey for locations and staff training for initial screenings. This task is expected to require the initial two months and include set-up and training, survey of utility partners and selection of sampling locations for preliminary screen, and preliminary screening of locations.

- Task 2: Initial sampling and GC/MS/O analysis. This task is expected to require the next 3 months and to comprise collection and analysis of samples for GC/MS/O and data analysis.
- Task 3: Report summary. This task is expected to require months six to nine months.

10.3.6 Schedule

For the purpose of scheduling, it has been assumed that notice to proceed for this research activity will be issued by May 1, 2007.

Table 10-6. Non-H_2S Odor- and Corrosion-Causing Compounds: Schedule.

	Start Date	End Date
Task 1	05/01/2007	07/01/2007
Task 2	07/01/2007	10/31/07
Task 3	10/31/07	04/31/08

10.3.7 Budget

Table 10-7. Non-H_2S Odor- and Corrosion-Causing Compounds: Budget.

	Total ($)
Task 1	37,904
Task 2	83,448
Task 3	85,096
Total	163,000

10.4 Effectiveness of Treatment Methods for Compounds Other Than Hydrogen Sulfide

Measurement and odor treatment techniques are usually designed for hydrogen sulfide, but numerous, less-well-understood compounds contribute to odor and warrant attention. Since treatment technologies are geared to hydrogen sulfide, other compounds are sometimes less effectively treated and can represent the bulk of odors remaining after treatment.

10.4.1 Problem Statement

Neglecting the spectrum of odor-causing compounds has led in some cases to failure of odor control systems even while hydrogen sulfide is effectively treated. Likewise, the success of many odor control systems may depend on the absence of other compounds in significant concentrations. A study is needed to evaluate how effectively odor compounds, other than hydrogen sulfide, are controlled with existing treatment systems.

10.4.2 Proposed Research Gap Evaluation Method

10.4.2.1 Objective

The objective of the propose study is to determine the effectiveness of existing odor control technologies for removing odor compounds other than hydrogen sulfide.

10.4.2.2 Program

The program of research would consist of the following three phases: 1) facility selection, 2) facility sampling, and 3) results documentation.

Phase 1—Facility Selection During Phase 1, up to 20 wastewater collection facilities where odor control technologies are in place will be identified. The facilities where permission can be obtained to carry out sampling should be selected. Also, a range of odor control technologies should be represented among the 20. Three months will be allocated for Phase 1.

Phase 2—Facility Sampling Air samples will be collected at each facility upstream and downstream of the control technology. GC-MS will be used to identify VOCs. Additionally, reduced sulfur compound analysis will be used to identify the dominant reduced sulfur compounds in each sample, hydrogen sulfide among them. Odor panel analyses will be used to determine the character and concentration change from upstream to downstream each treatment facility. Six months will be budgeted for Phase 2.

Phase 3—Results Documentation During Phase 3 results measured during Phase 2 will be documented. A report will be produced that provides guidance on the effectiveness of existing treatment technologies for controlling odor compounds. Three months will be provided to complete Phase 3.

10.4.3 Schedule and Budget

The cost of Phase 1 should consist mostly of labor required to research facilities that would best lend themselves to this project and to contact those utilities. Table 10-8 summarizes the schedule and budget.

Table 10-8. Effectiveness of Treatment Methods for Compounds Other Than Hydrogen Sulfide: Schedule and Budget

	Schedule	Budget ($)
Phase 1	05//07–08/07	17,500
Phase 2	08/07–02/08	97,000
Phase 3	02/08–05/08	37,400
Total	12 months	151,900

10.5 Life-Cycle Analysis of Chemical Addition on Downstream Processes

Liquid phase treatment of wastewater to remove dissolved sulfides usually involves adding significant quantities of chemicals. Several classes of chemical are used. Chemical oxidants (hydrogen peroxide, sodium hypochlorite) are used to oxidize sulfide to sulfate or sulfur. Iron salts are used to precipitate sulfide to an insoluble solid form. Nitrate is added to prevent sulfide from forming. pH adjustors such as magnesium hydroxide are used to prevent sulfide from entering the gas phase.

Many the chemicals used in liquid phase treatment may have side effects on the downstream treatment plant. For example, chlorine used to oxidize sulfide may have a residual disinfectant effect on activated sludge processes. Iron salts can cause increased sludge production. Nitrate addition may increase the nitrogen load on the plant. Some chemical suppliers claim that their products have beneficial side effects on downstream processes. For example, a supplier of magnesium hydroxide claims that its product improves flocculation in the primary clarifier.

10.5.1 Problem Statement

Although effects of chemical addition on downstream processes have been reported anecdotally, little quantitative information is available on these effects. In particular, the life-cycle analysis (LCA) of protecting infrastructure and preventing emissions verses downstream wastewater treatment impacts are unknown. A study is needed to quantify the downstream effects of chemical addition to the collection system and compare their performance using an LCA framework that calculates technical options based on economic, social, and environmental parameters. LCA is an ISO-standardized framework for the comparison of alternative products or services from an environmental perspective. The fundamental idea behind LCA is much broader than life-cycle costing (LCC) and consists of an assessment of all the physical and energetic flows that occur as a consequence of the selection of a particular process (see Figure 10-1).

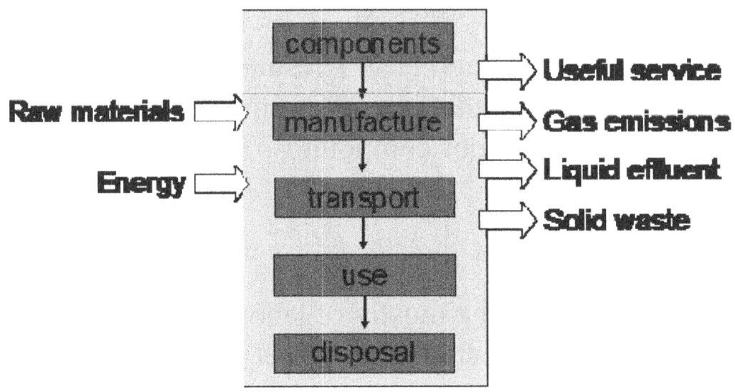

Figure 10-1. Example Process upon Which LCA Can Be Based.

10.5.2 Proposed Research Gap Evaluation Method

10.5.2.1 Objectives

The proposed study has the following objectives: 1) Identify effects of chemicals commonly added to collection systems on downstream processes, 2) develop LCA framework, and 3) quantify effects of chemical dosing on wastewater treatment using LCA.

10.5.2.2 Program

The program of research would consist of the following three phases: 1) Utility survey, 2) utility records analysis, and 3) LCA analysis.

Phase 1—Utility Survey In Phase 1, a survey will be conducted of utilities where collection system chemical addition is used. The observations of plant operators of downstream effects due to upstream chemical addition will be recorded. Plant records, where available, will be compiled for a period of at least one year before chemical addition was implemented until at least one year after. Four months will be budgeted to complete Phase 1.

Phase 2—Utility Records Analysis During Phase 2, records will systematized into a form whereby relevant treatment plant operating parameters can be compared before and after the start of upstream chemical addition. Before/after differences in operating parameters (such as aeration rate and sludge production) will be tracked and normalized for external conditions parameters (such as wastewater temperature and flow rate). Trends in operation parameter changes will be identified for each upstream chemical addition method. Trends will be analyzed to determine whether they can be attributed to upstream chemical addition. Where strong evidences suggest a relationship between upstream chemical additions and a downstream effect, LCA analysis will be used to compare performance (Phase 3). Four months will be budgeted to complete Phase 2.

Phase 3—LCA Analysis In Phase 3, an LCA framework will be developed to compare the performance of chemical dosing on collection system operation and wastewater treatment. This will involve dialogue with a Reference Panel of industry stakeholders to help define the sustainability criteria and relative significance of different key performance indicators. These quantitative key performance indicators will be used for the technology assessment. As a minimum, LCA requires 1) a clear definition the process being examined; 2) the compilation of an inventory of the flows in and out of the process and the product or products it yields; 3) characterisation and aggregation of these flows into environmental themes and impact assessment using models suitable for assessment of global and regional environmental impacts. This process will take into account financial and social analyses, and enable decision makers to determine the relative sustainability of alternative options. Data collected during Phase 2 will be evaluated and reported. Eight months will be budgeted to complete Phase 3.

10.5.3 Budget

Table 10-9 summarizes the schedule and budget.

Table 10-9. Life-Cycle Analysis of Chemical Addition on Downstream Processes: Schedule and Budget.

	Schedule	Budget ($)
Phase 1	4 months	TBD
Phase 2	4 months	TBD
Phase 3	4 months	TBD
Total	12 months	TBD

10.6 Sulfide Generation Rates in Wastewater Collection Systems: Relative Importance of Substrates

10.6.1 Objective

The objective of this research is to gain a better understanding of the biochemical and ecological factors that influence the rate at which aqueous sulfide is generated in collection systems. This would lead to a better understanding of the variation of sulfide levels in different sewers and enable improvements in existing models for predicting sulfide accumulation in sewer networks. The results of the proposed experiments may enable more judicious application of sulfide suppressing treatments such as calcium nitrate and liquid oxygen.

10.6.2 Rationale

Models to predict sulfide generation rates in wastewater collection systems have been proposed by Pomeroy and Hvitved-Jacobsen. In both models, the sulfide generation rate has been correlated to a general measure of the wastewater's organic strength. In the Pomeroy model, the sulfide generation rate is described as a function of BOD. Hvitved-Jacobsen bases the generation rate on the square root of the quantity (COD-50). Although the literature identifies a broad range of substrates that support the metabolism of sulfate-reducing bacteria, it also suggests that the primary substrates are restricted to a group of short-chain fatty acids and alcohols. Sulfate reducers appear to occupy a niche near the end of the metabolic food chain and, as such, depend upon other organisms to break down larger organic molecules. If this hypothesis is correct, then the two models describing sulfide generation are incomplete.

The above hypothesis makes the following predictions that are not considered by either the Pomeroy or Hvitved-Jacobsen formulations:

- Sulfide generation rates will be low in upstream sewer reaches because the substrates that have the strongest sulfidogenic potential will be present at only low levels.

- Sulfide generation rates will tend to increase as the wastewater flow moves down the collection system. The increase is a consequence of the degradation of poorly sulfidogenic substrates to metabolic intermediates preferred by sulfate-reducing bacteria. This assumes that the dissolved oxygen level in the wastewater remains at low concentrations that do not allow strongly sulfidogenic substrates to be cleared by biological oxidation, and that the flow has not been subject to significant dilution from other streams.
- The presence of dissolved oxygen in the upstream sewer reaches will delay the onset of sulfidogenesis. Dissolved oxygen will suppress the net production rate by the non-sulfate-reducing bacteria of relevant sulfidogenic substrates. Dissolved oxygen will also allow non-sulfate-reducing bacteria to compete for and clear the sulfidogenic substrates that are produced.
- The effect of an increase of dissolved oxygen in a wastewater stream already supporting sulfide generation will depend upon the level of sulfidogenic substrates at the point of increase the length of the reach over which dissolved oxygen remains elevated. When the wastewater in enriched in sulfidogenic materials, the effect of the dissolved oxygen increase may be minimal. If the dissolved oxygen increase occurs where sulfidogenic substrate is scarce, the available substrates may be cleared by biological oxidation. In this case, the dissolved oxygen increase may have effects that last a significant distance downstream.

Verification of these hypotheses would improve our understanding of the processes underlying the accumulation of sulfide in wastewater collection system. The accumulation of sulfide has direct impacts on the potential for odors to be emitted from sewers and on the corrosion of sewer crowns. Corrosion is a problem that plagues certain collection systems. The knowledge could be applied to target sulfide suppressing technologies to reaches where they will have maximal effect.

The following is proposed to assess the validity of these hypotheses on sulfidogenic substrates and oxygen levels and to identify most significant substrates with regard to sulfide generation:

- In conjunction with our utility partners, surveys of a sewer system will be conducted to identify areas where sulfide generation rates are low, intermediate, and high. Concrete coupons will be incubated at selected locations for the purpose of developing biofilms that have a microbial ecology representative of the various sulfidogenic conditions.
- The metabolic preferences of the sulfate-reducing population will be assessed in the laboratory. Coupons will be retrieved from the sewer and carefully washed to deplete them of substrates of sewer origin. The coupons will then be incubated in a mineral media containing one or more of the substrates preferred by sulfate-reducing bacteria. The substrates will include acetate, propionate, formate, fumarate, malate, lactate, ethanol, mixtures of fatty acids up to 16 carbons long, and perhaps others. The rate of sulfide generation at various concentrations of substrate will be measured.
- Biofilms will be incubated in a media consisting of important biomolecules such as simple sugars, amino acids, short polypeptides, polysaccharides, and mono-, di-, and triglycerides. The media, however, will be void of substrates known to be strongly sulfidogenic. The object of this experiment is to assess the hypothesis that larger biomolecules support sulfidogenesis only weakly. To supplement data collected when incubating biofilms in a well-defined media, the biofilms will also be incubated in wastewater collected from locations where sulfide generation is negligible and vigorous.

- The activity of sulfate-reducing bacteria is inhibited in the presence of dissolved oxygen. Oxygen concentrations in wastewater reflect the dynamic balance between reaeration processes and uptake by the biofilm. Increases in dissolved oxygen concentration along a sewer can be induced by changes in slope. The effect of dissolved oxygen changes on sulfide generation rates may be influenced by the time history of the wastewater before it enters the more oxygenated reach. The longer the wastewater has been in an anaerobic or microaerobic condition, the greater the buildup of sulfidogenic substrates, and the longer the time necessary to clear them out by biological oxidation. High concentration of sulfidogenic substrate may mitigate the inhibitory affects of the oxygen. It may also affect the time required for strongly sulfidogenic conditions to reestablish themselves following a sewer transition that results in a slowing of the reaeration rate. These hypothesized effects of dissolved oxygen can be evaluated using the coupon-grown biofilms.
- Should the proposed hypotheses correlating sulfide generation rates to the concentration be supported by the in-vitro studies proposed above, a follow-up study would be conducted to analyze for these substrates in wastewater collected from various locations. The object of this effort would be to obtain further confirmation of the mechanisms proposed above to explain variations in sulfide generation rates.

10.6.3 Duration and Budget

Total budget:	To be determined
In-kind contribution:	To be determined
WERF cost:	To be determined

10.7 Pure Oxygen Treatment of Gravity Sewers

Pure oxygen injection has been employed to prevent sulfide generation in force mains. Pressurized pipes flowing full allow much higher oxygen concentrations (and fewer injection stations) than can be applied in gravity sewers. Because of the high unit cost to transport oxygen and high capital costs to generate oxygen onsite, oxygen injection into gravity sewers is usually ruled out. However, small-scale pressure swing adsorption (PSA) could enable oxygen injection spaced frequently enough along gravity sewers to prevent anaerobic conditions from developing between injection points.

10.7.1 Problem Statement

If oxygen-generating equipment could be made small enough, oxygen injection could save costs by replacing chemical addition in gravity sewers. The feasibility of developing small PSA units for frequent spacing along gravity sewers should be studied.

10.7.2 Proposed Research Gap Evaluation Method

10.7.2.1 Objectives

The objectives of the proposed study are the following:

- Calculate the range of oxygen generation capacity appropriate for controlling sulfide generation in 0.5-mgd and larger gravity sewers.
- Identify technical and market obstacles preventing PSA manufacturers from developing small-scale PSA systems.
- Determine the feasibility of introducing a small-scale PSA system to the odor control market.

10.7.2.2 Program

The program of research will consist of a feasibility study to achieve the research objectives.

- **Task 1**—Determine the most appropriate sizes for an onsite oxygen generator to be used for injecting oxygen into gravity sewers.
- **Task 2**—Determine the capital cost, above which oxygen injection would not compare favorably to chemical addition.
- **Task 3**—Seek a partnership with a manufacturer of PSA systems, and determine the feasibility of developing a system of the size and cost needed to compete with chemical addition.
- **Task 4**—Develop a market analysis for presentation to manufacturers should Tasks 1 through 3 indicate a favorable market opportunity for small-scale PSA.

The intention is that demonstrating a favorable opportunity for oxygen generation manufacturers would foster the development of systems that could save a lot of money for utilities. For example, Orange County (Calif.) Sanitation District already spends $3 million per year on chemical addition and has calculated that oxygen would cost one-fifteenth of that if it could be delivered with the right equipment. At least eight months should be budgeted for this project.

10.7.3 Budget

Total budget:	To be determined
In-kind contribution:	To be determined
WERF cost:	To be determined

10.8 Liquid Phase Biological Treatment

Methods that seek to modify microbial metabolism, change microbial populations, or otherwise affect the microbes responsible for generating odors in wastewater collection systems are broadly classified as liquid phase biological treatment methods. These methods are distinct from liquid phase chemical treatment methods, which seek to chemically react with odor-causing compounds and convert them to nonodorous species.

Three main types of biological treatment methods have been marketed and applied with varying degrees of success. One method, bioaugmentation, seeks to deluge a collection system with bacteria specially bred to oxidize sulfide as it is generated, thereby reversing the natural anaerobic conversion of sulfate to sulfide. Another method, enzyme blocking, uses substances to interrupt the bacterial metabolic process that converts sulfate to sulfide. The third method, nitrate addition, seeks to supply sulfate-reducing bacteria with an alternate electron acceptor, whereby bacteria convert nitrate to nitrogen gas instead of converting sulfate to sulfide.

10.8.1 Problem Statement

The effectiveness of nitrate addition is well documented both in practice and in the technical literature, but the effectiveness of the other two biological treatment methods has not been sufficiently verified or documented. In some reports, the effectiveness of biological methods (other than nitrate) was inconclusive (Onondaga County Department of Drainage and Sanitation, 2001). In others, biological treatment methods were tested concurrently with nitrate or chemical methods, and the contribution of bioaugmentation or enzyme blocking could not be determined (Arthur and Anker, 2000; Kim et al., 2005). In some studies, anecdotal reports of the effectiveness of biological treatment methods are provided without showing data or system characteristics (Sercombe, 1995).

Because of prolific patenting and marketing of bioaugmentation and enzyme-blocking products, there is need for systematic research verifying the effectiveness of these biological methods. An independent study is needed to test leading biological methods under realistic conditions, without the influence of other methods. All relevant parameters should be reported, and the duration of the study should allow sufficient time to arrive at conclusive results.

10.8.2 Proposed Research Gap Evaluation Method

10.8.2.1 Objectives

The objectives of the proposed study are the following:

- Identify leading biological treatment methods (excluding nitrate addition, which has been sufficiently tested).
- Determine conclusively the level of effectiveness of each leading product.
- Determine the applications and doses for products determined to be effective.

10.8.2.2 Program

The program of research would consist of the following three phases: 1) study design, 2) pilot-scale testing, and 3) results documentation.

Phase 1—Study Design Phase 1 will be implemented in order to plan a study that can produce successful and conclusive results. During Phase 1, three months will be budgeted to complete the following three tasks:

- **Task 1**—A survey will be conducted to identify the biological treatment methods most commonly used and most likely to prove effective. Two bioaugmentation products and two enzyme-blocking products will be selected for testing.
- **Task 2**—A representative sewer system will be identified that can provide conditions for successful trials of the products to be tested. The sewer system should convey municipal wastewater, have long gravity sewers and force mains, generate significant concentrations of hydrogen sulfide and other odors, and have no existing liquid phase treatment program.
- **Task 3**—A pilot-scale testing regime will be designed. Appropriate doses of each of the four products will be selected; a feed system will be designed; feed durations will be selected; measurement methods will be selected; parameter variations will be selected; and a quality control program will be developed.

Phase 2—Pilot-Scale Testing During Phase 2 the products identified in Task 1 will be tested sequentially according to the protocol developed in Task 2. Phase 2 will consist of one task for each of the four products (Tasks 4 through 8). Eight months (two for each product) will be budgeted to complete Phase 2. The eight months will be divided between two warm seasons. One bioaugmentation method and one enzyme blocker will be tested during the first warm season; the other two methods will be tested during the following warm season. In this way, enough time will be allowed that the effects of the first method do not interfere with the following methods. Also, each method will be tested under comparable conditions.

Phase 3—Results Documentation During Phase 3, results measured during Phase 2 will be documented. In Task 9, a report will be generated that presents testing methods, parameter measurements, results, and conclusions. In Task 10 publication will be sought for the Task 9 report. Three months will be budgeted for Phase 3. The project will be considered complete upon the submission of an article to a major technical journal.

10.8.3 Budget

Table 10-10 summarizes the schedule and budget.

Table 10-10. Liquid Phase Biological Treatment: Schedule and Budget.

	Schedule	Budget
Phase 1	3 months	TBD
Phase 2	8 months	TBD
Phase 3	3 months	TBD
Total	14 months	TBD

10.9 Magnesium Hydroxide Odor Control Mechanisms Other Than pH Adjustment

Magnesium hydroxide ($Mg(OH)_2$) is a common chemical used prevent sulfide release in sewers. It works primarily as a pH adjuster, raising wastewater pH to between 8.5 and 9. In this pH range sulfide tends to remain in solution, where it does not cause odor and corrosion problems. Magnesium hydroxide is less soluble than other pH adjusters, such as caustic soda and lime. As such, it buffers the pH rather than reacting quickly, a favorable characteristic.

The pH adjustment effects of magnesium hydroxide are generally accepted in engineering practice, but other less understood benefits have been observed and attributed to the chemical. Premier Chemical, a supplier of magnesium hydroxide, claims that sulfide complexes with Mg^{2+} ions, forming a suspension that keeps its form during treatment processes. This action purportedly is similar to sulfur precipitation by iron salts and results in sulfide removal and pH adjustment. Premier also claims that Mg^{2+} catalyzes the oxidation of dissolved sulfides. The rate of sulfide oxidation is believed to increase two to three times in the presence of magnesium ion.

10.9.1 Problem Statement

Although there is anecdotal evidence of sulfide removal by magnesium hydroxide, independent research has not been carried out to verify the claims of chemical suppliers. An independent study is needed to establish the mechanisms responsible for sulfide removal by magnesium hydroxide and to quantify dosages that can be used to design chemical treatment systems.

10.9.2 Proposed Research Gap Evaluation Method

10.9.2.1 Objectives

The objectives of the proposed study are the following:

- Determine experimentally chemical mechanisms responsible for sulfide removal by magnesium hydroxide.
- Verify or debunk existing theories for sulfide removal mechanisms.
- Quantify dosage rates for sulfide removal by magnesium hydroxide.

10.9.2.2 Program

The program of research would consist of the following three phases: 1) study design, 2) experimentation, 3) results documentation.

Phase 1—Study Design Phase 1 will be implemented in order to plan a study that can produce successful and conclusive results. During Phase 1, existing mechanisms will be evaluated and experiments designed that can demonstrate whether or not the mechanisms are valid. Three months will be budgeted to complete Phase 1.

Phase 2—Experimentation An experimentation phase will be carried out during Phase 2. The program will depend on methods determined in Phase 1. Six months will be budgeted to complete this phase.

Phase 3—Results Documentation During Phase 3 results measured during Phase 2 will be documented. A report will be generated that presents testing methods, parameter measurements, results, and conclusions. Three months will be budgeted for Phase 3.

10.9.3 Budget

Table 10-11 summarizes the schedule and budget.

Table 10-11. Magnesium Hydroxide Odor Control Mechanisms Other Than pH Adjustment: Schedule and Budget.

	Schedule	Budget
Phase 1	3 months	TBD
Phase 2	6 months	TBD
Phase 3	3 months	TBD
Total	12 months	TBD

10.10 Design Basis and Standards for Air Jumpers

Air jumpers are used to convey foul air from sewer line headspace at siphons and other sewer headspace bottlenecks. Pressurization, and thus off-gassing, occurs without them or with them if they are undersized. Air jumpers usually are designed by estimating the volume of air needed to maintain negative pressure on the upstream end.

10.10.1 Problem Statement

Orange County (Calif.) Sanitation District has encountered odor issues associated with inadequately sized air jumpers. There is a general lack of technical literature on design standards and guidelines for sizing air jumpers, and thus a study is needed to develop a set of industry guidelines for designing air jumpers.

10.10.2 Proposed Research Gap Evaluation Method

10.10.2.1 Objective

The objective of the proposed study is to develop a document that will serve as a guideline for application of sewer air jumpers.

10.10.2.2 Program

The research program will consist of establishing the basis for air jumpers. The end product will be a comprehensive document outlining the theory and methodology for sizing air jumpers.

10.10.3 Budget

Table 10-12 summarizes the schedule and budget.

Table 10-12. Design Basis and Standards for Air Jumpers: Schedule and Budget.

Schedule	Budget
6 months	TBD

10.11 Effectiveness of Biotechnology in Cold Climates

The effectiveness of biotechnology is temperature dependent within the normal range of operating temperatures. Below that range, odor compound removal efficiency can be quite sensitive to temperature. Shareefdeen et al. (2004) studied biofilter removal of hydrogen sulfide at low temperatures in a laboratory and found that removal began to decrease precipitously below 5°C. Moreover, cold weather can cause inlet air to lose humidity, because of condensation, before entering the media. This hampers the effectiveness of biofilters.

10.11.1 Problem Statement

Since cold temperatures hamper biotechnology effectiveness, full-scale biotechnology systems should be evaluated for their effectiveness in cold weather environments. The study should identify locations where odor control functions properly during warm weather but fails during the winter. These cases should be compared to similar cases where the systems do not fail. Also, the effectiveness of modifications such as insulation and steam addition should be evaluated.

10.11.2 Proposed Research Gap Evaluation Method

10.11.2.1 Objectives

The objectives of the proposed study are the following:

- Determine the effectiveness of biotechnology during cold weather in the field.
- Identify design characteristics that permit successful cold weather operation
- Identify system changes that can improve cold weather operation

10.11.2.2 Program

The program of research will consist of the following three phases: 1) facility identification, 2) facility testing, and 3) results documentation.

Phase 1—Facility Identification Phase 1 will be implemented in order to locate cold weather facilities that use biotechnology for odor control. Four facilities that function poorly during cold weather but well the rest of the time will be sought. Another four facilities that function well during cold *and* warm weather will be sought. Permission must be obtained to study the odor control systems and to make temporary process changes. Three months will be budgeted for Phase 1.

Phase 2—Facility Testing During Phase 2, the eight facilities will be tested during two seasons. Warm weather testing will be carried out to verify that each facility operates well during the warm months. Cold weather testing will be carried out to evaluate each system's performance during cold weather. Differences between well-functioning and poorly functioning systems will be evaluated. The most promising system changes will be tried on the poorly functioning systems. Two months will be budgeted for warm weather testing and four for cold weather testing.

Phase 3—Results Documentation During Phase 3 a report will be produced documenting the results of Phases 1 and 2. The report will contain guidance for designing biotechnology systems for cold weather odor control. Three months will be budgeted for Phase 3.

10.11.3 Budget

Table 10-13 summarizes the schedule and budget.

Table 10-13. Effectiveness of Biotechnology in Cold Climates: Schedule and Budget.

	Schedule	Budget
Phase 1	3 months	TBD
Phase 2	6 months	TBD
Phase 3	3 months	TBD
Total	12 months	TBD

10.12 Fat, Oil, and Grease Odors in Sewers

Fats, oils, and grease (FOG) in wastewater are known to cause odor problems in sewers. However, most research on odors in sewers has focused on sulfur compounds and nitrogenous compounds. FOG forms a separate phase on the wastewater surface and may require control measures different than those for other odor-causing constituents.

10.12.1 Problem Statement

Research is needed to determine which liquid phase treatment methods are effective in treating odors associated with FOG. The study should use existing literature as a starting point and provide a practical basis for addressing odor problems associated with FOG.

10.12.2 Research Proposal

10.12.2.1 Objective

The objective of the proposed research is to determine the effectiveness of various liquid phase chemical and biological treatment methods for removing or preventing odors associated with FOG.

10.12.2.2 Program

The program of research will consist of three phases: 1) Identify FOG odor locations, 2) conduct bench-scale testing, and 3) prepare practical guidance manual.

Phase 1—Identify FOG Odor Locations Phase 1 will consist of a survey of collection systems known to emit odors caused by FOG. Wastewater from locations with severe FOG odors will be selected for bench-scale studies.

Phase 2—Conduct Bench-Scale Testing In Phase 2, wastewater samples from the four most severe FOG odor locations determined in Phase 1 will be tested using liquid phase odor control technologies. Bench-scale testing of the wastewater will be conducted using the most likely control chemicals, as found in the technical literature. Four months will be budgeted for Phase 2.

Phase 3—Prepare Practical Guidance Manual In Phase 3, a practical guidance manual will be prepared. The manual will contain the information gained in Phases 1 and 2. Phase 3 will be budgeted for 2 months.

10.12.3 Budget

Table 10-14 summarizes the schedule and budget.

Table 10-14. Fat, Oil, and Grease Odors in Sewers: Schedule and Budget.

	Schedule	Budget
Phase 1	3 months	TBD
Phase 2	4 months	TBD
Phase 3	3 Months	TBD
Total	10 months	TBD

10.13 Sewers as Process Reactors

Wastewater collection systems are designed and operated to serve as conveyance networks only, but sewers actually function as reactors, where numerous chemical and biological transformations occur within wastewater during transport. There appears to be great potential to harness those reactions to serve as part of the overall wastewater treatment process. The dual use of sewers for conveyance and pretreatment could add capacity or reduce costs of downstream wastewater treatment.

10.13.1 Problem Statement

Though research has been conducted to understand transformation reactions within sewers, they have not been harnessed to achieve specific treatment goals or to alter the nature of odorous and corrosive gas generation in sewers. A study is needed that attempts to design a dual-purpose wastewater conveyance/treatment system. Successful implementation of such a design could pave the way for a paradigm shift from the view that sewers are conveyance systems containing problematic reactions to the view that sewers are pretreatment process reactors.

10.13.2 Proposed Research Gap Evaluation Method

10.13.2.1 Objectives

The proposed research has the following objectives: 1) Identify wastewater treatment goals that could most feasibly be achieved; 2) create a retrofit design for a generic sewer system; and 3) offer the design for implementation in an existing system.

10.13.2.2 Program

The program of research would consist of three phases corresponding to objectives listed above: 1) Identify treatment goals, 2) retrofit design, and 3) design implementation.

Phase 1—Identify Treatment Goals In Phase 1 a study of the existing literature will be carried out to identify treatment processes that either benefit or decrease load on downstream treatment processes and that can be achieved within a sewer. Relevant potential design information will then be compiled from the technical literature for use in Phase 2. Three months will be budgeted for Phase 1.

Phase 2—Retrofit Design In Phase 2, a design will be created to retrofit a sewer system with typical or generic characteristics to add changes or equipment that will enable the system to function as a treatment reactor. The retrofit will be designed to meet the treatment goals determined in Phase 1. Parameters for the generic system will be selected with existing systems in mind. Five months will be allotted to complete Phase 2.

Phase 3—Design Implementation In Phase 3 the design and rationale developed in Phases 1 and 2 will be presented to selected owners in an attempt to sell the retrofit design. The design could be provided free of charge in exchange for the opportunity to have it implemented in a real system. Efforts will be made to generate interest in the project among prospective owners before starting. Phase 3 will be ongoing until agreement is reached with a collection system owner.

10.13.3 Budget

Table 10-15 summarizes the schedule and budget.

Table 10-15. Sewers as Process Reactors: Schedule and Budget.

	Schedule	Budget
Phase 1	3 months	TBD
Phase 2	5 months	TBD
Phase 3	TBD	TBD
Total	8+ months	TBD

APPENDIX A

PUBLISHED LITERATURE KEYWORD LIST

Search categories (not to be confused with review categories):

1. Collection Components and Corrosion
    a. Review categories: 1, 3, 4 (from phase 1 outline)
    b. Search team: LACSD
2. Containment, Ventilation, and Gas-Phase Treatment
    a. Review categories: 7, 8 (from phase 1 outline)
    b. Search team: OCSD
3. Liquid Phase and Biological Treatment (Enzymes, Bugs, etc.)
    a. Review categories: 5, 6 (from phase 1 outline)
    b. Review team: Stuetz and Cesca
4. Emissions Modeling and Compounds
    a. Review categories 2, 9 (from phase 1 outline)
    b. Review team: Corsi and Ward

Group 1

["Wastewater collection" OR Odor OR Sewer* OR "Collection System*"] AND
"Pump station"
"Force Main"
Gravity AND
    Sewer OR Interceptor OR Line
Siphon
"Wet well"
Manhole
Interceptor
"Air release valve"
"Chemical feed"
"Odor control equipment"
"Oil water separator"
Drop
corrosion
Liner*

Coating
Pig*
Tunnels
"CSO/SSO system*
Hydrogen sulfide
Sulfides
Air jumper
Corrosion resistant concrete
Grease trap

Group 2

["Wastewater collection" OR Odor OR Sewer* OR "Collection System*"] AND
Ventilation
- "Odor containment"
- Duct*
- "air handler"
- Headspace
- Cover*
- "Odor capture"
- "Ventilation rate"
- "Air exchange"

"Wet scrub*"
- "Packed Tower*"
- "Mist scrub*"
- Carbon
- Media
- Adsorption
- Absorption
- Biofilter
- Biotower
- Thermal
- Natural Ventilation
- Thermal
- Natural ventilation
- Forced ventilation
- corrosion

Group 3

["Wastewater collection" OR Odor OR Sewer* OR "Collection System*"] AND
Hydrogen Peroxide
- Chorine
- Hypochlorite
- Iron
- "sulfur precipitation"
- Nitrate
- "Liquid phase"

Oxidation
Permanganate
Chlorite
Ferrous
Ferric
Caustic
Hydroxide"Anaerobic layer"
Biofilm
"Slime layer"
Anthraquinone
"Biological treatment"
Bacteria* AND
Additives OR inhibitors OR anaerobic OR "sulfur reducing"
"Caustic shocking"
Corrosion

Group 4

["Wastewater collection" OR Odor OR Sewer* OR "Collection System*"] AND
Wastewater AND Emissions AND Models
INTERCEPTOR
BASTE
WATER9
TOXCHEM
(other models)
"Sulfide generation"
"Odor models"
"Hydrogen sulfide emissions"
"Reduced sulfur"
"Odor panel"
"D/T"
ammonia
mercaptan
sulfide
corrosion rate
sulfide formation

APPENDIX B

GRAY LITERATURE SOLICITATION LETTER AND UTILITY PARTNER LIST

May 10, 2007

Utility or POTW Manager
Water Reclamation Facility
One Breezy Lane
Anytown, Your State 12345

Subject: Water Environment Research Foundation (WERF)
Minimization of Odors and Related Corrosion in Collection Systems (WERF Project #04-CTS-1)

Dear Utility Manager:

Before you is a timely and important request to support WERF by providing "gray" literature information related to odors and corrosion from your collection systems.

Odors that create nuisance conditions and citizen complaints often originate in wastewater collection systems because of their proximity to the public. As a result, odor impacts from wastewater collection systems have been targeted recently by many national and local newspapers, citizens groups, and regulatory agencies. Whether there are regulatory implications or not, most wastewater utilities recognize that reduction of odors from their operations is simply good policy. Corrosion related to odor emissions can also cause collection system failures and result in costly impacts to capital improvement programs. WERF has initiated a study focused on the prevention and mitigation of odors and related corrosion in collection systems. This study is intended to provide useful operational and design guidance to collection systems owners and operators.

As such, we request your consideration in providing important information of potential interest to WERF in connection with the referenced project. The first phase of the project will include a comprehensive assessment of recent technical literature that addresses the sources and control of wastewater collection system odors and corrosion. The research team is using several methods to find and identify pertinent, "published" literature, but the team also recognizes some important literature and information on the subject may exist in the form of unpublished reports, many of them prepared for (or by) wastewater utilities such as yours. We refer to unpublished but important literature such as this by the term "gray literature."

The research team is interested in first obtaining information about the type and subject matter of gray literature that you have that addresses wastewater collection system odors and corrosion, then obtaining copies of any gray literature that the team feels is pertinent to the project. The team has developed a form for listing information about gray literature that you may have. The form is attached. We would appreciate if you could take a few minutes to complete the attached form and return it.

In order to stay on schedule, the project team is soliciting the participation and support of your organization with this gray literature request by December 16, 2005. Please fill out the attached form and either email or fax it to Dirk Apgar, Principal co-investigator from King County , whose email address is dirk.apgar@metrokc.gov and fax no. is 206-684-1945.

Mr. Apgar is responsible for collecting the survey forms and forwarding them to the WERF research team for compilation and review. He will then be back in touch with you or your designee for the purpose of obtaining copies of gray literature designated by the research team as being of interest to this project.

As with all WERF research efforts, the timeliness of responses to requests for information such as this is critical to timely completion of the research reports for the benefit of the WERF subscribers and the wastewater industry as a whole.

If you need further information regarding the technical background of the study feel free to contact Dirk Apgar or the project team leader Chris Easter at ceaster@ch2m.com, or call him at 804-559-4993.

Thank you in advance for your cooperation.

Best regards,

Dirk Apgar
Co-Principal Investigator
King County

Gray Literature Summary Reply			
Study Title	Date of Work	Authorship	Brief Description of Information Provided

APPENDIX C

VENDOR LIST

Vendor	Contact
Praxair	Layana Moris
Aire Liquide	Philip Bost
Tridon Chemical	Elizabeth Cattana
Kemiron (formerly Eaglebrook)	Scott Selsethal
Bay Chemical	Doyle Keller
General Chemical	Karen Ruehl
Basic Chemical (formerly Vulcan)	Jerry Nemis
Davis Process/US Filter	Patrick Laidlaw
Carus Chemical (Permanganate)	John Boll
Premiere Chemical (Thioguard)	Matt Madolora
Advance Oxidation (VTX)	Mr. Cox
Altivia (Nitrazyme, and VX-456)	Jennifer Miller
FMC Chemicals	Marketing Division
US Peroxide (peroxide, PRI-SC)	John Walton
Filter Inovations Inc. (BioStreme, Hydralogic)	Irene
Byo-Gon*	Gary Sober
Ortec	Antonio Gisbert, CEO
Novozymes Biologicals	Greg DeLozier and Bernie
Enzymatic Odor Solutions, Inc. (EOSI)	Mike Durham & Wolfram Pinker
Drew Industrial (Ashland Chemical)	Marketing Division

APPENDIX D

LITERATURE REVIEW FORM

WERF Article Review Form

Tracking Number		
Author(s)		
Year		
Title		
Journal or Conference		
Reviewer		
Topic Statement		Enter a sentence Highlighting Main Topic
Topics Categories:		List, in order of importance, topic category numbers of topics addressed by the paper
Quality Rating		Rate quality A, B, or C
Important Conclusions:		List important findings or conclusions of the paper.
Presents Experimental Data?		Enter 'Yes' or 'No'
Presents Model Results?		Enter 'Yes' or 'No'
Presents Case Study(ies)?		Enter 'Yes' or 'No'
Identified Data Gaps:		List potential topics or areas that are underrepresented in the technical literature.

Topic Category Number	
1	Collection System Component Physical Design Guidelines
2	Odor and Corrosion Compound Formation Mechanisms
3	Corrosion Mechanisms
4	Corrosion Protection including linings and coatings
5	Liquid Phase Chemical Treatment
6	Liquid Phase Biological Treatment including Enzymes and bacterial seeding
7	Ventilation Approaches
8	Gas Phase Treatment Approaches
9	Emissions corrosion models
10	Pump Stations and force mains
11	Gravity Sewers
12	Tunnels
13	CSO/SSO systems
14	Misc. (Junction Boxes, Drop Shafts, Siphons, etc.)

Quality Rating	
A	Highly applicable pertinent paper published in peer reviewed journal and provides sufficient data to backup conclusions. Essentially this is a high priority valuable paper with an excellent rating.
B	Applicable pertinent paper providing sufficient data to backup conclusions but does not have to be from a peer reviewed journal (i.e. could be from a conference proceeding, or well documented gray literature). Essentially this is a valuable paper with valuable information and a good rating.
C	Paper that either proves not to be pertinent to our areas of interest or that does not provide backup information for major conclusions. Essentially this is a poor quality paper that either does not apply to our interest or has questionable conclusions.

References

Arthur, P.J.; Anker, L.S. Hydrogen Sulfide Odor Control Technology. Tappi Paper Makers Conference and Trade Fair, Vancouver, BC, Canada, 2000, 1, 1–2.

ASCE. *Sulfide in Wastewater Collection and Treatment Systems*. Manuals and Reports on Engineering Practice No. 69. Reston, VA. 1989.

Ashley, R.; Bertrand-Krajewski, J.L.; Hvitved-Jacobsen, T. Sewer Solids—20 Years of Investigation. *Water Sci. Technol.* 2005, *52*(3), 73–84.

Badia, J.; Chen, C.; Esfandi, E.; Kimbell, W. Caustic Spray for Sewer Crown Corosion Control. County Sanitation District of Los Angeles, CA. 1992.

Beeldens, A.; Monteny, J.; Vincke, E.; De Belie, N.; Van Gemert, D.; Taerwe, L.; Verstraete, W.; Beeldens, A. Resistance to Biogenic Sulphuric Acid Corrosion of Polymer-Modified Mortars. *Cem. Concr. Compos.* 2001, *23*(1), 47–56.

Bell, L.W.; Shook, W.E.; Norris, T. Mitigating the Corrosion of Concrete Pipe and Manholes. In *Concrete Pipe for the New Millennium*; Kaspar, Enyart, Eds.; Special Technical Publication 1368, ASTM International: West Conshohocken, PA, 2000; pp 3–10.

Bentzen, G.; Smith, A.T.; Bennett, D.; Webster, N.J.; Reinholt, F.; Sletholt, E.; Hobson, J. Controlled Dosing of Nitrate for Prevention of H_2S in a Sewer Network and the Effects on the Subsequent Treatment Processes. *Water Sci. Technol.* 1995, *31*(7), 293.

Bielefeldt, A.R.; Lutz, M.; Taylor, S.; Song, M.; Wyeno, L.; Anderson, J. Use of Iron Chemicals for Odor Control: Lab Studies and Full Scale Results. Presented at WEFTEC.02: 75th Annual Water Environment Federation Annual Technical Exhibition and Conference, Chicago, IL, Sept. 28–Oct. 2, 2002.

Boehm, M.; Devinny, J.; Jahani, F.; Rosen, G. On a Moving-Boundary System Modeling Corrosion in Sewer Pipes. *Appl. Math Comput.* 1998, *92*(2–3), 247–269.

Boon, A.G.; Vincent, A.J.; Boon, K.G. Avoiding the Problems of Septic Sewage. *Water Sci. Technol.* 1998, *37*(1), 223–231.

Bourgeois W.; Gardey G.; Servieres M.; Stuetz R.M. A Chemical Sensor Array–Based System for Protecting Wastewater Treatment Plants. *Sensors Actuators B* 2003, *91*, 109–116.

Chern, J.M.; Yang, S.P. Measuring and Modeling of Oxygen Transfer Rate in a Drop Structure. *Ind. Eng. Chem. Res.* 2004, *43*(23), 7657–7663.

Cho, K.-S.; Ohtani, T.; Mori, T. A Newly Isolated Fungus, Sporormia Concretivora, Participates in the Corrosion of Concrete Sewer Pipes. Presented at the International Specialized Conference on the Sewer as a Physical, Chemical, and Biological Reactor, Aalborg, Denmark, May 16–18, 1994.

Churchill, P.; Elmer, D. Hydrogen Sulfide Odor Control in Wastewater Collection Systems. *J. New England Water Environ. Assoc.* 1999, *33*(1), 57.

Converse, B.M.; Schroeder, E.D.; Iranpour, R.; Cox, H.H.J.; Deshusses, M.A. Odor and Volatile Organic Compound Removal from Wastewater Treatment Plant Headworks Ventilation Air Using a Biofilter. *Water Environ. Res.* 2003, *75*(5), 444–454.

Corsi, R.L.; Quigley, C.; Melcer, H.; Bell, J. Aromatic VOC Emmissions from a Municipal Sewer Interceptor. *Water Sci. Technol.* 1995, *31*(7), 137–145.

Coughenour, J.R., II; Walz, T.; Wokulich, P.; Vera, S.; Phillips, J.; Husband, J. Using Centrate Nitrification to Reduce Odors at the 91st Avenue WWTP. Presented at WEFTEC '03: 76th Annual Water Environment Federation Annual Technical Exhibition and Conference, Los Angeles, CA, Oct. 11–15, 2003.

Crowley, J.; Concha, L. Fix This Smell Now—Quick Response Odor Control Options. Presented at the Odors and Toxic Air Emissions Conference, Albuquerque, NM, Apr. 28–May 1, 2002.

Davidson, S.; Green, J.; Mann, J.; Lamb, E. Assessing the Effectiveness of Sewer Headspace Foul Air Extraction. Presented at WEFTEC '02: 75th Annual Water Environment Federation Annual Technical Exhibition and Conference, Chicago, IL, Sept. 28–Oct. 2, 2002.

De Belie, N.; Monteny, J.; Beeldens, A.; Vincke, E.; Van Gemert, D.; Verstraete, W. Experimental Research and Prediction of the Effect of Chemical and Biogenic Sulfuric Acid on Different Types of Commercially Produced Concrete Sewer Pipes. *Cem. Concr. Res.* 2004, *34*(12), 2223–2236.

Delgado, S.; Alvarez, M.; Rodriguez-Gomez, L.E.; Elmaleh, S. Transportation of Reclaimed Wastewater through a Long Pipe: Inhibition of Sulphide Production by Nitrite from the Secondary Treatment. *Environ. Technol.* 2004, *25*(3), 365–371.

Devai, I.; Delaune, R.D. Effectiveness of Selected Chemicals for Controlling Emission of Malodorous Sulfur Gases in Sewage Sludge. *Environ. Technol.* 2002, *23*(3), 319–329.

Dewaters, J.E.; Zander, A.K.; Grimberg, S.J. Use of Ferric Salts for Controlling Sulfide Odors in High-Strength Pulp and Paper Manufacturing Wastes. *Environ. Eng. Sci.* 1999, 16(6), 441.

Easter, C.; Quigley, C.; Burrowes, P.; Witherspoon, J.; Apgar, D. Biotechnology Gains Ground as a Tool for Odor and Air Emissions Control. *Water Environ. Technol.* 2005, *17*(5), 27–32.

Edwards, M.; Courtney, B.; Heppler, P.S.; Hernandez, M. Beneficial Discharge of Iron Coagulation Sludge to Sewers. *J. Environ. Eng.* 1997, *123*(10), 1027.

Edwini-Bonsu, S. Air Flow in Sanitary Sewer Systems: A Physically Based Approach. Ph.D. Thesis, University of Alberta, Alberta, Canada, 2004.

Edwini-Bonsu, S.; Steffler, P.M. Air Flow in Sanitary Sewer Conduits due to Wastewater Drag: A Computational Fluid Dynamics Approach. *J. Environ. Eng. Sci.* 2004, *3*(5), 331–342.

Einarsen, A.M.; Aesoy, A.; Rasmussen, A.I.; Bungum, S.; Sveberg, M. Biological Prevention and Removal of Hydrogen Sulphide in Sludge at Lillehammer Wastewater Treatment Plant. *Water Sci. Technol.* 2000, *41*(6), 175–182.

Fenner, R.A.; Stuetz, R.M. The Application of Electronic Nose Technology to Environmental Monitoring of Water and Wastewater Treatment Activities. *Water Environ. Res.* 1999, *71*(3), 282–289.

Fong, J.C. Sulfide Control with Hydrogen Peroxide in Large Diameter Sewers, Los Angeles County Sanitation Districts, CA, 1986.

Frechen, F.-B.; Köster, W. Odour Emission Capacity of Wastewaters—Standardization of Method and Application. *Water Sci. Technol.* 1998, *38*(3), 61–69.

Gebler, S.H.; Powers, L.; Willems, T.; Detwiler, R.; Gebler, S.H. Petrographic Examination Used to Analyze a Distressed Sewer Line. *J. Prot. Coat. Linings* 2002, *19*(5), 49–52.

George Butler Associates, Inc. 1999 Wastewater Infrastructure Improvements Odor/Corrosion Control Full-Scale Pilot Study Final Report, City of Lee's Summit, Missouri, Lee's Summit, Missouri.

Ghosn, S.G.; Cox, H.W. Full Scale Pilot Study Using the VTX Process for the Removal of Sulfide at the Virginia Initiative Plant. Presented at WEF/AWMA Air and Odor Specialty Conference, Hartford, CT. April, 2006.

Gostelow, P.; Parsons, S.A. Sewage Treatment Works Odour Measurements. *Water Sci. Technol.* 2000, *41*, 33–40.

Groleau, J.; Clausen, T.; Hunniford, D.; Ellis, K.; Huza, S. Bioxide® Improves Air Quality, Controls Odors and Corrosive Gas Generation in Dewatered Sludge. TAPPI Press, International Environmental Conference, Montreal, QC, Canada, 2002, 681–684.

Guan, S.W., Corrosion Protection by Coatings for Water and Wastewater Pipelines. 46th Appalachian Undergraduate Corrosion Short Corse, West Virginia University, PA, May, 2001.

Gudjonsson, G.; Vollertsen, J.; Hvitved-Jacobsen, T. Dissolved Oxygen in Gravity Sewers—Measurement and Simulation. *Water Sci. Technol.* 2002, *45*(3), 35–44.

Hagekhalil, A.; Jenson, B.; Hamilton, J.; Joyce, J.; Kemmerle, R. Control of Hydrogen Sulfide and Related Odor and Corrosion Problems in the City of Los Angeles Wastewater Collection System Utilizing Nitrate Addition: A Biochemical Process. Presented at Water Environment Federation Specialty Conference on Odor and Volatile Organic Compound Emission Control for Municipal and Industrial Wastewater Treatment Facilities, Jacksonville, FL. 1994.

Hamilton, J.; Kemmerle, R.; Wada, F.; Jacobs, P.; Langley, J.; Crosse, J.; Ludwig, K.; Hydrogen Sulfide Abatement in the City of Los Angeles Wastewater Collection System. Presented at Water Environment Federation Specialty Conference on Odor and Volatile Organic Compound Emission Control for Municipal and Industrial Wastewater Treatment Facilities, Jacksonville, FL, 1992.

Harshman, V. Follow Your Nose. *Water Environ. Technol.* 2003, *15*(7), 30–35.

Hewayde, E.; Allouche, E.N.; Nakhla, G.F. Experimental Investigations of the Effect of Selected Admixtures on the Resistance of Concrete to Sulfuric Acid Attack. Presented at New Pipeline Technology, Security and Safety, Baltimore, MD, July 13–16, 2003.

Hobson, J. The Odour Potential: A New Tool for Odour Management. *J.CIWEM,Water Environ. J.* 1995, *9*, 458-463.

Hobson, J. and Yang., G. The Ability of Selected Chemicals for Suppressing Odor Development in Rising Mains. *Water Sci. Technol.* 2000, *41*(6), 165–173.

Huang, Y.; Ortiz, L.; Garcia, J.; Aguirre, P.; Mujeriego, R.; Bayona, J.M. Use of Headspace Solid-Phase Microextraction to Characterize Odour Compounds in Subsurface Flow Constructed Wetland for Wastewater Treatment. *Water Sci. Technol.* 2004, *49*(9), 89–98.

Huisman, J.L.; Gasser, T.; Gienal, C.; Kuehni, M.; Krebs, P. Quantification of Oxygen Fluxes in a Long Gravity Sewer. *Water Res.* 2004, *38*(5), 1237–1247.

Hvitved-Jacobsen, T. *Sewer Processes: Microbial and Chemical Process Engineering of Sewer Networks,* CRC Press, Boca Raton, FL. 2002. 237 pages.

Hvitved-Jacobsen, T.; Vollertsen, J. Odour Formation in Sewers. In *Odours in Wastewater Treatment Measurement, Modelling and Control: Odour Formation in Sewers*. Richard Stuetz and Franz-Bernd Frechen, eds. IWA Publishing, London, UK, 2001. 33-68.

Hvitved-Jacobsen, T.; Vollertsen, J.; Nielsen, P.H. A Process and Model Concept for Microbial Wastewater Transformation in Gravity Sewers. *Water Sci. Technol.* 1998a, *37*(1), 233–241.

Hvitved-Jacobsen, T.; Vollertsen, J.; Tanaka, N. Wastewater Quality Changes during Transport into Sewers—An Integrated Anaerobic Model Concept for Carbon and Sulfur Microbial Transformations. *Water Sci. Technol.* 1998b, *38*(10), 257–264.

Hvitved-Jacobsen, T.; Vollertsen, J. Tanaka, N. Wastewater Quality Changes During Transport in Sewers—An Integrated Aerobic and Anaerobic Model Concept for Carbon and Sulfur Microbial Transformations. *Water Sci. Technol.* 1999, *38*(10), 257–264.

Hvitved-Jacobsen, T.; Vollertsen, J.; Tanaka, N. Integrated Aerobic/Anaerobic Approach for Prediction of Sulfide Formation in Sewers. *Water Sci. Technol.* 2000, *41*(6), 107–116.

Ito, T.; Nielsen, J.L.; Okabe, S.; Watanabe, Y.; Nielsen, P.H. Phylogenetic Identification and Substrate Uptake Patterns of Sulfate-Reducing Bacteria Inhabiting an Oxic-Anoxic Sewer Biofilm Determined by Combining Microautoradiography and Fluorescent In Situ Hybridization. *Appl. Environ. Microbiol.* 2002, *68*(1), 356–364.

Jameel, P. The Use of Ferrous Chloride to Control Dissolved Sulfides in Interceptor Sewers. *J. Water Pollut. Control Fed.* 1989, *61*(2), 230–236.

James, K.J.; Stack, M.A. Rapid Determination of Volatile Organic Compounds in Environmentally Hazardous Wastewaters Using Solid Phase Microextraction. *Fresenius J. Anal. Chem.* 1997, *358*, 833–837.

Jefferson, B.; Hurst, A.; Stuetz, R.; Parsons, S.A. A. Comparison of Chemical Methods for the Control of Odors in Wastewater. *Process Safety Environ. Prot.* 2002, *80*(B2), 93–99.

Jin, Sulfide Control with Sodium Hyrdroxide in Large Diameter Sewers. Los Angeles County Sanitation Districts, Calif.1987.

Jordan, B.; Perkins, M.; Plummer, A.; Cleveland, P.; Tatum, B.; Woolsey, M. Pure Oxygen for Pure Air at Wastewater Treatment Plants. Presented at WEFTEC.03: 76th Annual Water Environment Federation Annual Technical Exhibition and Conference, Los Angeles, CA, Oct. 11–15, 2003.

Kahn, G.A. Continuous Caustic Addition for Sewer Corrosion Control. Los Angeles County Sanitation Districts. 1992.

Kienow, K.K.; Kienow, K.E. Evaluation of Sanitary Sewer Sulfide, Odor, and Corrosion Potential. In *Proceedings of the Pipeline Division Specialty Congress, August 1–4, 2004, San Diego, California*; Galleher, Jr., John J., Stift, Michael T., Eds; American Society of Civil Engineers: Reston, VA, 2004.

Kim, D.K. Effect of OrTec Biocatalyst on Biofilms. OrTec Ltd. 2004.

Kim, H.H.; McConnell, L.; Pamirez, M.; Abu-Orf, M.; Choi, H.L.; Peot, C. Characterization of Odors from Limed Biosolids Treated with Nitrate and Anthraquinone. *J. Environ. Sci. Health Part A Toxic Hazard. Subst. Environ. Eng.* 2005, *40*(1), 139–149.

Kim, H.; Murthy, S.; McConnell, L.L.; Peot, C.; Ramirez, M.; Strawn, M. Characterization of Wastewater and Solids Odors Using Solid Phase Microextraction at a Large Wastewater Treatment Plant. *Water Sci. Technol.* 2002, *46*(10), 9–16.

Kotronarou, A.; Hoffman, M.R. Peroxymonosulfate: An Alternative to Hydrogen Peroxide for the Control of Hydrogen Sulfide. *Res. J. Water Pollut. Control Fed.* 1991, *63*(7), 965–970.

Lahav, O.; Lu, Y.; Shavit, U.; Loewenthal, R.E. Modeling Hydrogen Sulfide Emission Rates in Gravity Sewage Collection Systems. *J. Environ. Eng.* 2004, *130*(11), 1382–1389.

Laing, D.G.; Eddy, A.; John, B.D. Perceptual Characteristics of Binary, Trinary, and Quaternary Odor Mixtures Consiststing of Unpleasant Constituents. *Physiol. Behav.* 1994, *56*(1), 81–93.

Lamb Technical Services. Major Interceptor Odor and Hydrogen Sulfide Monitoring Project. Pima County Wastewater Management Department, Tucson, AZ, 1998.

Little, B.J.; Ray, R.I.; Pope, R.K. Relationship between Corrosion and the Biological Sulfur Cycle: A Review. *Corrosion* 2000, *56*(4), 433–443.

Little, J.C.; McGinnis, D.F. Hypolimnetic Oxygenation: Predicting Performance using a Discrete-Bubble Model. *Water Sci. Technol. Water Supply* 2001, *1*(4), 185–191.

Lloyd, F.M. Controlling Reduced Sulfur Compounds. *Environ. Technol. Atlanta Ga.* 1997, *7*(5), 57.

Martos, P.A.; Pawliszyn, J. Calibration of Solid Phase Microextraction for Air Analyses Based on Physical Chemical Properties of the Coating. *Anal. Chem.* 1997, *69*, 206–215.

Matos, J.S.; Aires, C.M. Mathematical Modelling of Sulphides and Hydrogen Sulphide Gas Build-up in the Costa Do Estoril Sewerage System. *Water Sci. Technol.* 1995, *31*(7), 255–261.

McGinley, D.L.; McGinley, M.A. Understanding Odor Panels and Odor Evaluations. Proceedings of the 90th Annual Meeting of the Air and Waste Management Association, Toronto, Ontario, Canada, 1997.

McGinnis, D.F.; Little, J.C. Bubble Dynamics and Oxygen Transfer in a Speece Cone. *Water Sci. Technol.* 1998, *37*(2), 285–292.

Melcer, H.; Bell, J.P.; Thompson, D.J.; Yendt, C.M.; Kemp, J.; Steel, P. Modeling Volatile Organic Contaminants' Fate in Wastewater Treatment Plants. *J. Environ. Eng.* 1994, *120*(3), 588–609.

Melbourne Water. *Hydrogen Sulphide Control Manual.* Melbourne, Victoria, Australia, 1989.

Michiels, A.; Salgaonkar, J.A. Computer Model for Estimation of Atmospheric Hydrogen Sulfide Concentrations in Sanitary Sewers. Presented at Water Environment Federation Specialty

Conference on Odor and VOC Organic Compound Emission Control for Municipal and Industrial Wastewater Treatment Facilities, Jacksonville, FL, April 24–27, 1994.

Mohleji, S.C.; Occiano, V.Y.; Wong, P.; Schafer, P.L. Odor Characterization and Control by Ferrous Chloride Addition in Sewers (Full-Scale Field Test). Presented at Water Environement Federation 65th Annual Conference and Exposition, New Orleans, LA, September 20–24, 1992.

Monteith, H.; Bell, J.; Harvey, T. Assessment of Factors Controlling Sewer Ventilation Rates. Presented at Water Environment Federation Specialty Conference on Control of Odors and VOC Emissions, Houston, TX, April 20–23, 1997.

Monteny, J.; De Belie, N.; Vincke, E.; Verstraete, W.; Taerwe, L. Chemical and Microbiological Tests to Simulate Sulfuric Acid Corrosion of Polymer-Modified Concrete. *Cement Concr. Res.* 2001, *31*(9), 1359–1365.

Moody, T.C.; Riek, G.C. Sulfide Suppression. *Water Environ. Technol.* 1999, *11*(2), 45–50.

Moore, B.C. Downflow Bubble Contact Aeration Technology (Speece Cone) for Sediment Oxygenation. Remediation of Contaminated Sediments: Proceedings of the Second International Conference on Remediation of Contaminated Sediments Venice, Italy, October 3, 2003a, 903–910.

Moore, L.W. Need Relief. *Water Environ. Technol.* 2003b, *15*(10), 32–35.

Mori, T.; Koga, M.; Hikosaka, Y.; Nonaka, T.; Mishina, F.; Sakai, Y.; Koizumi, J. Microbial Corrosion of Concrete Sewer Pipes, H_2S Production from Sediments and Determination of Corrosion Rate. *Water Sci. Technol.* 1991, *23*, 1275–1282.

Mori, T.; Nonaka, T.; Tazaki, K.; Koga, M.; Hikosaka, Y.; Noda, S. Interactions of Nutrients, Moisture and pH on Microbial Corrosion of Concrete Sewer Pipes. *Water Res.* 1992. *16*(1), 29–37.

Nica, D.; Davis, J.L.; Kirby, L.; Zuo, G.; Roberts, D.J. Isolation and Characterization of Microorganisms Involved in the Biodeterioration of Concrete in Sewers. *Int. Biodeterior. Biodegrad.* 2000, *46*(1), 61–68.

Nielsen, P.H. Sulfur Sources for Hydrogen Sulfide Production in Biofilm from Sewer Systems. *Water Sci. Technol.* 1991, *23*, 1265–1274.

Nielsen, P.H.; Raunkjaer, K.; Norsker, N.H.; Jensen, N.A.; Hvitved-Jacobsen, T. Transformations Wastewater in Sewer Systems—A Review. *Water Scienc Technol.* 1992, *25*(6), 17–31.

Nielsen, P.H.; Raunkjaer, K.; Hvitved-Jacobsen, T. Sulfide Production and Wastewater Quality in Pressure Mains. *Water Sci. Technol.* 1998, *37*(1), 97–104.

Nielsen, A.H.; Vollertsen, J.; Hvitved-Jacobsen, T. Chemical Sulfide Oxidation of Wastewater—Effects of pH and Temperature. *Water Sci. Technol.* 2004, *50*(4), 185–192.

Nielsen, A.H.; Lens, P.; Vollertsen, J.; Hvitved-Jacobsen, T. Sulfide-Iron Interactions in Domestic Wastewater from a Gravity Sewer. *Water Res.* 2005a, *39*(12), 2747–2755.

Nielsen, A.H.; Hvitved-Jacobsen, T.; Vollertsen, J. Kinetics and Stoichiometry of Sulfide Oxidation by Sewer Biofilms. *Water Res.* 2005b, *39*(17), 4119–4125.

Norsker, N.H.; Nielsen, P.H.; Hvitved-Jacobsen, T. Influence of Oxygen on Biofilm Growth and Potential Sulfate Reduction in Gravity Sewer Biofilm. Presented at the International Specialized Conference on the Sewer as a Physical, Chemical, and Biological Reactor, Aalborg, Denmark, May 16–18, 1994.

Nowak, R.; Rupp, F.; Schattovits, G. Well-proven solutions for problems with bad odour emission from Vienna's sewer system. Presented at WEF Odors and Air Emissions 2004, Bellevue, WA, April 18–21, 2004.

O'Brien, J.E.; Fuller, R.D.; O'Hare, J.P. Hydro-Vac Treatment of Septage for Odor Elimination and Increased Revenue. *J. New Engl. Water Environ. Assoc.* 1994, *28*(2), 136–145.

Okabe, S.; Ito, T.; Satoh, H.; Watanabe, Y. Effect of Nitrite and Nitrate on Biogenic Sulfide Production in Sewer Biofilms Determined by the Use of Microelectrodes. *Water Sci. Technol.* 2003, *47*(11), 281–288.

Olmstead and Hamlin. Converting Portions of the Los Angeles Outfall Sewer into a Septic Tank. *Eng. News*. 1900, *44*, 317–318.

Olson, D.A., Rajagopalan, S., and Corsi, R.L., Ventilation of Sewers: The Role of Thermal Gradients, *Adv. Environ. Res.* 1997a, *1*(3), 312–322.

Olson, D.A., Rajagopalan, S., and Corsi, R.L., Ventilation of Industrial Process Drains: Mechanisms and Effects on VOC Emissions, *J. Environ. Eng. Am. Soc. Civ. Eng.* 1997b, *123*(9): 939–947.

Onondaga County Department of Drainage and Sanitation. Oneida Lakeshore Sanitary District: Enzybate-E Demonstration Trial., Syracuse, NY, 2001.

Orange County Sanitation District. H_2S Control Testing: Summary of Thioguard® Magnesium Hydroxide Results. Orange County, CA, 2000.

OrTec. Application of Biocatalyst for Controlling Emission of Odors, 2002.

OrTec. Application of Biocatalysts for Controlling Ammonia and Hydrogen Sulfide Emissions, 2002.

Padival, N.A.; Kimbell, W.A.; Redner, J.A. Use of Iron Salts to Control Dissolved Sulfide in Trunk Sewers. *J. Environ. Eng.* 1995, *121*(11), 824.

Padival, N. The Use of Iron Salts to Control Dissolved Sulfide in Districts' Trunk Sewers. Los Angeles County Sanitation Districts, CA, 1993.

Pai, T.Y.; Ouyang, C.F.; Liao, Y.C.; Leu, H.G. Oxygen Transfer in Gravity Flow Sewers. *Water Sci. Technol.* 2000, *42*(3), 417–422.

Parker, W.; Ryan, H.A. Tracer Study of Headspace Ventilation in a Collector Sewer *J. Air Waste Manage. Assoc.* 2001, *54*(4), 582–592.

Perry, S.A.L.; Gravette, B.; Einfeld, B.; Postma, L. Odor control for the City of Everett, Washington wastewater collection system. Presented at WEFTEC '97 Annual Conference and Exposition, Chicago, IL, Oct. 18–22, 1997.

Pescod, M.B.; Price, A.C. Fundamentals of Sewer Ventilation as Applied to the Tyneside Sewerage Scheme, *Water Pollut. Control*, 1981, *90*(1), 14–33.

Pomeroy, R.D. Generation and Control of Sulfide in Filled Pipes. *Sew. Ind. Wastes* 1959. *31*(9), 1082–1095.

Pomeroy, R.D. *Process Design Manual for Sulfide Control in Sanitary Sewerage Systems*, NTIS No. PB-260479. EPA, Washington, DC, 1974.

Pomeroy, R.D.; Parkhurst, J.D. The Forecasting of Sulfide Build-up Rates in Sewers. *Prog. Water Technol.* 1977. *9*(3), 621–628.

Pope, R.J. Collection System Odors—Going with the Flow. Presented at Air and Waste Management Association's 89th Annual Meeting, 1996.

Pope, R.J.; Pohren, R.L.; Warren, R.J. Deep Tunnel Mysteries—Solving Odor/Corrosion Problems. Presented at Air and Waste Management Association's 91st Annual Meeting, San Diego, CA, June 14–18, 1998.

Prewitt, R.H. The Use of Locally Generated Oxygen to Control Force Main Odors. Presented at WEF Odors and VOC Emissions 2000, Cincinnati, OH, April 16–19, 2000.

Quigley, C.; Easter, C.; Burrowes, P.; Witherspoon, J. Biotechnology-Based Odour Control: Design Criteria and Performance Data. *Water Sci. Technol.* 2004, *50*(4), 319–326.

Rafson, H.J. *Odor and VOC Control Handbook*. McGraw-Hill, Columbus, OH. 1998.

Richard, D.; Zimmerman, R.; Johnson, T. The Control of Sulfides by Chemical Addition of Ferrous Chloride Saves Fargo (ND) System. Presented at Water Environment Federation's 66th Annual Conference, October 3–7, 1993.

Rodriguez-Gomez, L.E.; Delgado, S.; Alvarez, M.; Elmaleh, S. Inhibition of Sulfide Generation in a Reclaimed Wastewater Pipe by Nitrate Dosage and Denitrification Kinetics. *Water Environ. Res.* 2005, *77*(2), 193–198.

Saricimen, H.; Shameem, M.; Barry, M.S.; Brahim, M.; Abbasi, T.A. Durability of Proprietary Cementitious Materials for Use in Wastewater Transport Systems. *Cem. Concr. Compos.* 2003, *25*(4–5), 421–427.

Sato, H.; Hirose, T.; Kimura, T.; Moriyama, T.; Nakashima, Y. Analysis of Malodorous Volatile Substances of Human Waste: Feces and Urine. *J. Health Sci.* 2001, *47*(5), 483–490.

Sato, H.; Morimatsu, H.; Kimura, T.; Moriyama, Y.; Yamashita, T.; Nakashima, Y. Analysis of Malodorous Substances of Human Feces. *J. Health Sci.* 2002, *48*(2), 179–185.

Schmidt, R.; Regan, M.; Grieb, R.; Witherspoon, J.; Greene, B. Innovative Feedback Control System for Chemical Dosing to Control Treatment Plant Odors. Presented at WEFTEC 2000: 73rd Annual Water Environment Federation Annual Technical Exhibition and Conference, Anaheim, CA, Oct. 14–18, 2000.

Sercombe, D.C.W. The Control of Septicity and Odours in Sewerage Systems and at Sewage Treatment Works Operated by Anglian Water Services Limited. Presented at the International Specialized Conference on the Sewer as a Physical, Chemical, and Biological Reactor, Aalborg, Denmark, May 16–18, 1994.

Sercombe, D.C.W. The Control of Septicity and Odours in Sewerage Systems and at Sewage Treatment Works Operated by anglian Water Services Limited. *Water Sci. Technol.* 1995, *31*(7), 283.

Shand, M. Magnesium Hydroxide—An Alternative to Lime and Caustic Soda.

Shand, M.; Miller, T.; Manson, I. Reduction of Odors and Corrosion in Magnesia Coated Sewers.Presented at WEFTEC 1996, 69th Annual Conference and Exposition, Dallas, TX, Oct. 5–9, 1996.

Shareefdeen, Z.; Herner, B.; Webb, D.; Wilson, S.; Morris, A. Biofiltration of Hydrogen Sulphide (H_2S) in Cold Climates. Proceedings of the Air and Waste Management Association's Annual Meeting and Exhibition Indianapolis, IN. 2004, 1523-1530.

Sharma, V.K. Potassium Ferrate(VI): An Environmentally Friendly Oxidant. *Adv. Environ. Res.* 2002, *6*(2), 143.

Sima, H. Selective Inhibition of Acidophilic Thiobacilli for Application of Controlling Microbially Induced Corrosion in Concrete Sewers, Ph.D. Thesis, University of Arizona, Tucson, 1992.

Slezak, L.; So, S.; Jones, T.; Merry, C. Ferrous Chloride Addition Demonstration Program for WEF Specialty Conference on Odor and Volatile Organic Compound Emission Control for Municipal and Industrial Wastewater Treatment Facilities, Vancouver, British Columbia. April 24–27, 1994.

Smith, M.M.; Klunzinger, D. Dynamics of Sanitary Sewer Ventilation and Odor Release in the Heights/Hilltop and Soutwest/West Leg Interceptor Sewer Systems. Presented at WEFTEC 2001: 74th Annual Water Environment Federation Annual Technical Exhibition and Conference, Atlanta, GA, Oct. 13–17, 2001.

Soloducha, F.E. Operational Experience with a Force Main Air Injection System for Control of Hydrogen Sulfide Gas. Washington Suburban Sanitary Commission, Washington, DC, 1989.

Sorenson, H.; Joyce, J.; Day, D.; Fallara, C.T. Odor Control for Large Diameter Deep Sewer Tunnels—The City of Columbus Ohio. Presented at Water Environment Federation Conference on Odors and VOC Emissions, 2000.

Stuetz, R.M.; Engin, G.; Fenner, R.A. Sewage Odour Measurements Using a Sensory Panel and an Electronic Nose. *Water Sci. Technol.* 1998, *38*(3), 331–335.

Stuetz, R.M.; Fenner, R.A.; Hall, S.J.; Stratful, I.; Loke, D. Monitoring of Wastewater Odors Using an Electronic Nose. *Water Sci. Technol.* 2000, *41*(6, Control and Prevention of Odours in the Water Industry), 41–47.

Stuetz, R.M.; Bourgeois, W. Application of Non-Specific Gas Sensors for Detecting Odour Abatement Process Failure. *VDI Ber.* 2004, *1850*, 463–471.

Sye, W.-F.; Chen, W.-Y. Determination of Odorous Sulfur Compounds in Air by Tenax GR Preconcentration and Capillary Gas Chromatography in Combination with Flameless Sulfur Chemiluminescence Detection. *Toxicol. Environ. Chem.* 2000, *77*(3–4), 207–218.

Tanaka, N.; Takenaka, K. Control of Hydrogen Sulfide and Degradation of Organic Matter by Air Injection into a Wastewater Force Main. Presented at the International Specialized Conference on the Sewer as a Physical, Chemical, and Biological Reactor, Aalborg, Denmark, May 16–18, 1994.

Tanaka, N.; Hvitved-Jacobsen, T. Transformations of Wastewater Organic Matter in Sewers under Changing Aerobic/Anaerobic Conditions. *Water Sci. Technol.* 1998, *37*(1), 105–113.

Tanaka, N.; Hvitved-Jacobsen, T.; Ochi, T.; Sato, N. Aerobic-Anaerobic Microbial Wastewater Transformations and Re-aeration in an Air-Injected Pressure Sewer. *Water Environ. Res.* 2000, *72*(6), 665–674.

Tator, K.B. Preventing Hydrogen Sulfide and Microbiologically Influenced Corrosion in Wastewater Facilities. *Mater. Perform.* 2003, *42*(7), 32–37.

Trypus, J.F.; Gans, R.L.; Murhpy, T.J.; Doane, J.W. Evaluating and Solving Odor Problems Along a Large-Diameter Regional Interceptor Sewer. Presented at WEFTEC 2001: 74th Annual Water Environment Federation Annual Technical Exhibition and Conference, Atlanta, GA, Oct. 13–17, 2001.

Turkmen, M.; Dentel, S.K.; Chiu, P.C.; Hepner, S. Analysis of Sulfur and Nitrogen Odorants Using Solid-Phase Microextraction and GC-MS. *Water Sci. Technol.* 2004, *50*(4), 115–120.

Uhren, D.J.; Gilbert, G.W. Matrix Evaluation to Resolve Corrosion Risks for a Sewer Tunnel in Difficult Glacial Soils. Presented at Rapid Excavation and Tunneling Conference, Society of Mining, Metallurgy, and Exploration, Littleton, CO, 2001.

U.S. EPA. *Sulfide Control in Sanitary Sewerage Systems*. EPA 625/1-74-005. Washington, DC. 1974.

U.S. EPA. *EPA Design Manual for Odor and Corrosion Control in Sanitary Sewerage Systems and Treatment Plants*. EPA/625/1-85/018. Washington, DC, 1985.

US Peroxide. Collection System Sulfide Control Demonstration, Peroxide Regenerated Iron-Sulfide Control (PRI-SC), Phase-2: Miller Holder and Knott Interplant Trunk Lines, Laguna Niguel, CA, 2001.

Van Durme, G.P.; Berkenpas, K.J. Comparing Sulfide Control Products. *Operations Forum.* 1989, *6*(2) 12–19.

Vincke, E.; Boon, N.; Verstraete, W. Analysis of the Microbial Communities on Corroded Concrete Sewer Pipes—A Case Study. *Appl. Microbiol. Biotechnol.* 2001, *57*(5–6), 776–785.

Vinke, Elke; Van Wanseele, Ellen; Monteny, Joke; Beeldens, Anne; De Belie, Nele; Taerwe, Luc; Van Gemert, Dionys; Verstraete rug ac be, Willy Verstraete. Influence of Polymer Addition on Biogenic Sulfuric Acid Attack of Concrete. *Int. Biodeterior. Biodegrad.* 2002, *49*(4), 283.

Vollertsen, J.; Hvitved-Jacobsen, T. Resuspension and Oxygen Uptake of Sediments in Combined Sewers. *Urban Water* 2000, *2*(1), 21–27.

Vollertsen, J.; Hvitved-Jacobsen, T.; Nielsen, A.H. Stochastic Modeling of Chemical Oxygen Demand Transformations in Gravity Sewers. *Water Environ. Res.* 2005a, *77*(4), 331–339.

Vollertsen, J.; Nielsen, A.H.; Yang, W.; Hvitved-Jacobsen, T. Effects of in-Sewer Processes: A Stochastic Model Approach. *Water Scienc Technol.* 2005b, *52*(3), 171–180.

Vroom, A.H.; Aarsleff, L.; Vroom, C.H. Sulfur Concrete for Corrosion-Resistant Sewer Pipe. *ASTM Spec. Tech. Publ.* 2000, *1368,* 11–20.

Wahshat, T.M. Sulfur Mortar and Polymer Modified Sulfur Mortar Lining for Concrete Sewer Pipe, Ph.D. Thesis, Iowa State University, Ames, 2001.

Walsh, J.L. What's That Smell? *Ind. Wastewater* 1996, 4(33), 34–38.

Walton, J.R.; Velasco, M.S.; Ratledge, E. Peroxide Regenerated Iron-Sulfide Control (PRI-SC): Integrating Collection System Sulfide Control with Enhanced Primary Clarification by Adding Iron Salts and Hydrogen Peroxide. Presented at WEFTEC.03: 76th Annual Water Environment Federation Annual Technical Exhibition and Conference, Los Angeles, CA, Oct. 11–15, 2003.

Waltrip, D.; Snyder, E.G. Elimination of Odor at Six Major Wastewater Treatment Plants. *J. Water Pollut. Control Fed.* 1985, *57*(10), 1027–1032.

Watson, A.; Goggins, R.; Hunniford, D.; Overton, B.; Huza, S. Odor Control and Sulfide Removal from a Wastewater Treatment Plant Lagoon. Presented at TAPPI International Environmental, Health, and Safety Conference and Exhibit, Charlotte, NC, 2001.

Webster Environmental Associates, Inc. Oxygen Ionization System Demonstration Project. Orange County Sanitation District, CA, 2003.

WEF. *Odor Control in Wastewater Treatment Plants*. Manual of Practice 22. WEFPress, Alexandria, VA. 1995.

WEF. *Control of Odors and Emissions from Wastewater Treatment Plants*. Manual of Practice 25. WEFPress, Alexandria, VA. 2004.

WERF. *Identifying and Controlling Odor in the Municipal Wastewater Environment: Literature Search and Review.* Alexandria, VA. 2003.

Wilmot, P.D.; Cadee, K.; Kantinic, J.J.; Kavanagh, B.V. Kinetics of Sulfide Oxidation by Dissolved Oxygen. *J. Water Pollut. Control Fed.* 1988, *60*(7), 1264–1270.

Witherspoon, J.R.; Barnes, J.L. Comparison of Methods Used to Measure Odour at Wastewater Treatment Plant Fencelines. *VDI Ber.* 2004, *1850*, 363–373.

Witherspoon, J.; Allen, E.; Quigley, C. Modelling to Assist in Wastewater Collection System Odour and Corrosion Potential Evaluations. *Water Sci. Technol.* 2004, *50*(4), 177–183.

Witzgall, R.A.; Horner, I.S.; Schafer, P.L. Case Histories of Sulfide Corrosion: Problems and Treatments. *Water Environ. Technol.* 1990, 2(77), 42-47.

Wolf, T.A.; McManus, K. R.; Jennette, J. P. Odor Reduction in a Wastewater Conveyance Forcemain. Presented at WEF Odors and VOC Emissions 2000, Cincinnati, OH, April 16–19, 2000.

Wolstenholme, P.L.; Jabloner, M.L. Sulfide and Odor Control at Naval Station Everett. Presented at WEF Odors and Air Emissions 2004, Bellevue, WA, April 18–21, 2004.

Won, D.L. *Sulfide Control with Ferrous Chloride in Large-Diameter Sewers*. Los Angeles County Sanitation Districts, 1988.

Yamanaka, T.; Aso, I.; Togashi, S.; Tanigawa, M.; Shoji, K.; Watanabe, T.; Watanabe, N.; Maki, K.; Suzuki, H. Corrosion by Bacteria of Concrete in Sewerage Systems and Inhibitory Effects of Formates on Their Growth. *Water Res.*, 2002, *36*(10), 2636–2642.

Yongsiri, C.; Hvitved-Jacobsen, T.; Vollertsen, J.; Tanaka, N. Introducing the Emission Process of Hydrogen Sulfide to a Sewer Process Model (WATS). *Water Sci. Technol.* 2003a, *47*(4), 85–92.

Yongsiri, C.; Hvitved-Jacobsen, T.; Vollertsen, J.; Tanaka, N. Erratum: Introducing the Emission Process of Hydrogen Sulfide to a Sewer Process Model (WATS) (Water Science and Technology 47(4), 85–92). *Water Sci. Technol.* 2003b, *47*(12), 319–320.

Yongsiri, C.; Vollertsen, J.; Hvitved-Jacobsen, T. Hydrogen Sulfide Emission in Sewer Networks: A Two-Phase Modeling Approach to the Sulfur Cycle. *Water Sci. Technol.* 2004a, *50*(4), 161–168.

Yongsiri, C.; Vollertsen, J.; Hvitved-Jacobsen, T. Effect of Temperature on Air–Water Transfer of Hydrogen Sulfide. *J. Environ. Eng.* 2004b, *130*(1), 104–109.

Onondaga County Department of Drainage and Sanitation, Enzybate-E Demonstration Trial, Syracuse, NY. 2001.

Vinke, E.; Van Wanseele, E.; Monteny, J.; Beeldens, A.; De Belie, N.; Taerwe, L.; Van Gemert, D.; Verstraete, W. Influence of Polymer Addition on Biogenic Sulfuric Acid Attack of Concrete. Int. Biodeterior. Biodegrad. 2002, *49*(4), 283.

WASTEWATER UTILITY

Alabama
Montgomery Water Works & Sanitary Sewer Board

Alaska
Anchorage Water & Wastewater Utility

Arizona
Glendale, City of, Utilities Department
Mesa, City of
Peoria, City of
Phoenix Water Services Dept.
Pima County Wastewater Management
Safford, City of

Arkansas
Little Rock Wastewater Utility

California
Central Contra Costa Sanitary District
Corona, City of
Crestline Sanitation District
Delta Diablo Sanitation District
Dublin San Ramon Services District
East Bay Dischargers Authority
East Bay Municipal Utility District
Eastern Municipal Water District
El Dorado Irrigation District
Fairfield-Suisun Sewer District
Fresno Department of Public Utilities
Inland Empire Utilities Agency
Irvine Ranch Water District
Las Virgenes Municipal Water District
Livermore, City of
Los Angeles, City of
Los Angeles County, Sanitation Districts of
Napa Sanitation District
Orange County Sanitation District
Palo Alto, City of
Riverside, City of
Sacramento Regional County Sanitation District
San Diego Metropolitan Wastewater Department, City of
San Francisco, City & County of
San Jose, City of
Santa Barbara, City of
Santa Cruz, City of
Santa Rosa, City of
South Bayside System Authority
South Coast Water District
South Orange County Wastewater Authority
Stege Sanitary District
Sunnyvale, City of
Union Sanitary District
West Valley Sanitation District

Colorado
Aurora, City of
Boulder, City of
Greeley, City of
Littleton/Englewood Water Pollution Control Plant
Metro Wastewater Reclamation District, Denver

Connecticut
Greater New Haven WPCA
Stamford, City of

District of Columbia
District of Columbia Water & Sewer Authority

Florida
Broward, County of
Fort Lauderdale, City of
Miami-Dade Water & Sewer Authority
Orange County Utilities Department
Reedy Creek Improvement District
Seminole County Environmental Services
St. Petersburg, City of
Tallahassee, City of
Tampa, City of
Toho Water Authority
West Palm Beach, City of

Georgia
Atlanta Department of Watershed Management
Augusta, City of
Clayton County Water Authority
Cobb County Water System
Columbus Water Works
Fulton County
Gwinnett County Department of Public Utilities
Savannah, City of

Hawaii
Honolulu, City & County of

Idaho
Boise, City of

Illinois
American Bottoms Wastewater Treatment Plant
Greater Peoria Sanitary District
Kankakee River Metropolitan Agency
Metropolitan Water Reclamation District of Greater Chicago
Wheaton Sanitary District

Iowa
Ames, City of
Cedar Rapids Wastewater Facility
Des Moines, City of
Iowa City

Kansas
Johnson County Unified Wastewater Districts
Unified Government of Wyandotte County/ Kansas City, City of

Kentucky
Louisville & Jefferson County Metropolitan Sewer District
Sanitation District No. 1

Louisiana
Sewerage & Water Board of New Orleans

Maine
Bangor, City of
Portland Water District

Maryland
Anne Arundel County Bureau of Utility Operations
Howard County Department of Public Works
Washington Suburban Sanitary Commission

Massachusetts
Boston Water & Sewer Commission
Massachusetts Water Resources Authority (MWRA)
Upper Blackstone Water Pollution Abatement District

Michigan
Ann Arbor, City of
Detroit, City of
Holland Board of Public Works
Saginaw, City of
Wayne County Department of Environment
Wyoming, City of

Minnesota
Rochester, City of
Western Lake Superior Sanitary District

Missouri
Independence, City of
Kansas City Missouri Water Services Department
Little Blue Valley Sewer District
Metropolitan St. Louis Sewer District

Nebraska
Lincoln Wastewater System

Nevada
Henderson, City of
Las Vegas, City of
Reno, City of

New Jersey
Bergen County Utilities Authority
Ocean County Utilities Authority
Passaic Valley Sewerage Commissioners

New York
New York City Department of Environmental Protection

North Carolina
Charlotte/Mecklenburg Utilities
Durham, City of
Metropolitan Sewerage District of Buncombe County
Orange Water & Sewer Authority

Ohio
Akron, City of
Butler County Department of Environmental Services
Columbus, City of
Metropolitan Sewer District of Greater Cincinnati
Northeast Ohio Regional Sewer District
Summit, County of

Oklahoma
Oklahoma City Water & Wastewater Utility Department
Tulsa, City of

Oregon
Albany, City of
Clean Water Services
Eugene, City of
Gresham, City of
Portland, City of Bureau of Environmental Services
Water Environment Services

Pennsylvania
Hemlock Municipal Sewer Cooperative (HMSC)
Philadelphia, City of
University Area Joint Authority

South Carolina
Charleston Water System
Mount Pleasant Waterworks & Sewer Commission
Spartanburg Water

Tennessee
Cleveland, City of
Knoxville Utilities Board
Murfreesboro Water & Sewer Department
Nashville Metro Water Services

Texas
Austin, City of
Dallas Water Utilities
Denton, City of
El Paso Water Utilities
Fort Worth, City of
Houston, City of
San Antonio Water System
Trinity River Authority

Utah
Salt Lake City Corporation

Virginia
Alexandria Sanitation Authority

Arlington, County of
Fairfax County
Hampton Roads Sanitation District
Hanover, County of
Henrico, County of
Hopewell Regional Wastewater Treatment Facility
Loudoun County Sanitation Authority
Lynchburg Regional WWTP
Prince William County Service Authority
Richmond, City of
Rivanna Water & Sewer Authority

Washington
Everett, City of
King County Department of Natural Resources
Seattle Public Utilities
Sunnyside, Port of
Yakima, City of

Wisconsin
Green Bay Metro Sewerage District
Kenosha Water Utility
Madison Metropolitan Sewerage District
Milwaukee Metropolitan Sewerage District
Racine, City of
Sheboygan Regional Wastewater Treatment
Wausau Water Works

Australia
ACTEW (Ecowise)
South Australian Water Corporation
Sydney Water Corporation
Water Corporation of Western Australia

Canada
Lethbridge, City of
Regina, City of, Saskatchewan
Toronto, City of, Ontario
Winnipeg, City of, Manitoba

New Zealand
Watercare Services Limited

United Kingdom
United Utilities North West (UUNW)

STORMWATER UTILITY

California
Fresno Metropolitan Flood Control District
Los Angeles, City of, Department of Public Works
Monterey, City of
Sacramento, County of
San Francisco, City & County of
Santa Rosa, City of
Sunnyvale, City of

Colorado
Aurora, City of
Boulder, City of

Georgia
Griffin, City of

Iowa
Cedar Rapids Wastewater Facility
Des Moines, City of

Kansas
Overland Park, City of

Kentucky
Louisville & Jefferson County Metropolitan Sewer District

Maine
Portland Water District

North Carolina
Charlotte, City of, Stormwater Services

Pennsylvania
Philadelphia, City of

Tennessee
Chattanooga Stormwater Management

Texas
Harris County Flood Control District, Texas

Washington
Bellevue Utilities Department
Seattle Public Utilities

STATE

Arkansas Department of Environmental Quality
Connecticut Department of Environmental Protection
Kansas Department of Health & Environment
Kentucky Department of Environmental Protection
New England Interstate Water Pollution Control Commission (NEIWPCC)
Ohio River Valley Sanitation Commission
Urban Drainage & Flood Control District, CO

CORPORATE

ADS Environmental Services
Advanced Data Mining International
Alan Plummer & Associates
Alpine Technology Inc.
Aqua-Aerobic Systems Inc.
Aquateam–Norwegian Water Technology Centre A/S
ARCADIS
Associated Engineering
Black & Veatch
Blue Water Technologies, Inc.
Boyle Engineering Corporation
Brown & Caldwell
Burgess & Niple, Ltd.
Burns & McDonnell
CABE Associates Inc.
The Cadmus Group
Camp Dresser & McKee Inc.
Carollo Engineers Inc.
Carpenter Environmental Associates Inc.
CDS Technologies Inc.
CET Engineering Services
Chemtrac Systems Inc.
CH2M HILL
CRA Infrastructure & Engineering
CONTECH Stormwater Solutions
D&B/Guarino Engineers, LLC
Damon S. Williams Associates, LLC
Earth Tech Inc.
Ecovation
EMA Inc.
Environmental Operating Solutions, Inc.
Environ/The ADVENT Group, Inc.
Fay, Spofford, & Thorndike Inc.
Freese & Nichols Inc.
ftn Associates Inc.
Gannett Fleming Inc.
Garden & Associates, Ltd.
Geosyntec Consultants
GHD
Golder Associates Ltd.
Greeley and Hansen LLC
Hazen & Sawyer, P.C.
HDR Engineering Inc.
HNTB Corporation
Hydromantis Inc.
HydroQual Inc.
Infilco Degremont Inc.
Jacobson Satchell Consultants, Inc.
Jacques Whitford NAWE, Inc.
Jason Consultants LLC Inc.
Jordan, Jones, & Goulding Inc.
KCI Technologies Inc.
Kelly & Weaver, P.C.
Kennedy/Jenks Consultants
KMK Consultants
Komline Sanderson Engineering Corporation
Larry Walker Associates
Limno-Tech Inc.
The Low Impact Development Center Inc.
Malcolm Pirnie Inc.
Material Matters
McKim & Creed
Metcalf & Eddy Inc.
Monteco Corporation
MPR Engineering Corporation, Inc.
MWH
NewFields Water Resources, LLC
O'Brien & Gere Engineers Inc.
Odor & Corrosion Technology Consultants Inc.
Original Engineering Consultants, Ltd.
Oscar Larson & Associates
Parametrix Inc.
Parsons
Post, Buckley, Schuh & Jernigan
Praxair, Inc.
Ring Industrial Group
RMC Water & Environment
Ross & Associates Ltd.
Rothberg, Tamburini & Windsor, Inc.
SAIC
Savin Engineers
Siemens Water Technologies
Stantec Consulting Inc.
Stearns & Wheler, LLC
Stone Environmental Inc.
Stratus Consulting Inc.
Synagro Technologies Inc.
Tetra Tech Inc.
Trojan Technologies Inc.
Trussell Technologies, Inc.
URS Corporation
Wade-Trim Inc.
Westin Engineering Inc.
Weston Solutions Inc.
Woodard & Curran
Zenon Environmental Inc./GE Water Process Technology
Zoeller Pump Company

INDUSTRY

American Electric Power
American Water
ChevronTexaco Energy Research & Technology Company
The Coca-Cola Company
Dow Chemical Company
DuPont Company
Eastman Chemical Company
Eli Lilly & Company
Merck & Company Inc.
Premier Chemicals LLC
Procter & Gamble Company
Thames Water Utilities
Severn Trent Services Inc.
Suez Environnment
United Water Services LLC

Note: List as of 10/31/07

IWA Publishing's authorised EU representative for General Product Safety Regulations is Diane D'Arras, 15 rue Duret, 75116 Paris, France, e-mail: safety@iwap.co.uk.

Printed and bound by CPI Group (UK) Ltd, Croydon, CR0 4YY
17/09/2026
02218793-0006